The Human Question

What People Believe About Evolution, Human Origins, and the Beginning of Life

Hervey Cunningham Peoples

Manufactured in the United States of America

Publisher's Cataloging-in-Publication
(***Provided by Quality Books, Inc.***)

Peoples, Hervey Cunningham.
The human question : what people believe about evolution, human origins, and the beginning of life / Hervey Cunningham Peoples. -- 1st ed.
p. cm.
Includes bibliographical references and index.
LCCN 2002093494
ISBN 0-9722330-6-7

1. Human beings--Origin. 2. Human evolution.
I. Title.

GN281.P46 2003 599.93'8
QBI02-200643

Interior Design and Typesetting by Publishing Professionals, Port Richey, FL
Cover Design by George Foster

Published by Red Lion Press, P. O. Box 8850, Longboat Key, FL 34228

This work is dedicated
to my husband
David Alexander Peoples,
and to the memory of
my parents
Charles and Florence Cunningham.

"Come along with me, the best is yet to be."

Table of Contents

Acknowledgments

I have many people to thank.

I am indebted to the following individuals for their generous assistance and invaluable professional support during the research and writing of this book:

Dr. Jennifer A. Clack, Cambridge University Museum of Zoology

Dr. Richard M. Cornelius, Bryan College Archives

Dr. Francis B. Harrold, University of Nebraska-Kearney

Dr. Samuel M. Gon III, The Nature Conservancy of Hawaii

Richard Hammond, 3-D Artist

Dr. Susan C. Losh, Florida State University

Liesel and Jay Ritchie, Social Science Research Center, Mississippi State University

Dr. Charles H. Smith, Western Kentucky University

Dr. Paul Michael Taylor, Director, Asian Cultural History Program, Department of Anthropology, Smithsonian Institution

Dr. Christoph Zollikofer and Dr. Marie Ponce de Leon, Multi-Media Laboratory, University of Zurich, Switzerland.

I am grateful for the logistical support of the Department of Anthropology, Smithsonian Institution, Washington, D.C.; Natalie Moore of The Natural History Museum, London; Dave and Betty Lou LeDrew, Corner Brook, Newfoundland.

Finally, I wish to express my sincere appreciation to all those individuals throughout the past seven years who have so generously shared with me their private thoughts about where we came from. You know who you are. This book is for you.

Introduction

This is not a book about religion. This is a book about the variety of personal beliefs about human origins and the beginning of life—and the best evidence for each.

The Human Question brings a fresh approach to an age old question and a frank examination of all sides of the human origins debate. From evolution to "creation science" and reincarnation to life on other planets, we'll examine a wide variety of beliefs about human origins in the same way most people think of them—as related pieces of the human origins puzzle.

Only 10% of Americans say evolution alone is responsible for life on earth. Nearly 45% believe in the divine, supernatural creation of all life. Yet the private beliefs of these people are seldom black and white, and many have unanswered questions.

Over 100 million Americans, including 40% of scientists, have a strong belief in something they never talk about. These people believe in both evolution and a spiritual cause for human origins. This more interesting third voice is seldom heard above the polarizing din of the evolution/creation debate. How do these folks reconcile a belief based on two conflicting philosophies?

Several years ago I started casually asking people the not-so-casual question, "Where do you think we came from?" Most told me they had never been asked that question. Then came a breathless pause before they began to share with me their most intimate thoughts about life, death, science, faith, and the unknown.

Those initial interviews suggested that our notions about human origins and the beginning of life are far more complex than most national surveys suggest. I developed a Belief Survey to capture the unique aspects of these personal beliefs. The responses of over 1,000 Americans who participated in the Survey appear throughout the book.

More significantly, my Survey gave me access to a large number of people from whom I could solicit in-depth Personal

Interviews. Their opinions and reflections are the stories behind the statistics. For example, what we know and what we believe are often two very different things. More people believe in ESP and angels than in evolution. Excerpts from the Interviews appear in each chapter and reveal what we believe and why we believe it—often despite overwhelming evidence to the contrary.

Compelling scientific discoveries and innovative theories are painting a clearer picture of the development of life on Earth. But how and where did life begin?

This book unravels the web of scientific jargon and decodes the latest discoveries and current thinking. We'll clear up common misunderstandings and contradictory evidence, and come to some surprising conclusions.

For the first time we'll hear the voices of millions of folks who privately sense they hold a very different view of where we came from. You may recognize your own beliefs in their views and no longer feel, as one person told me, "as though I'm out here on my own believing in something so different from anything I have ever heard about that I need to start my own organization."

So come along with me as we travel into the realm of human belief in search of the answer to *The Human Question*.

Hervey Cunningham Peoples
August 11, 2002

The author can be contacted at:
P. O. Box 8850,
Longboat Key, FL 34228,
or via e-mail at *RedLionPress1@aol.com*
visit our Website: *www.HumanQuestion.com*

Chapter 1

The Million Dollar Question

The pursuit of human origins is intellectually demanding and rigorous. But it is more than that. Because the target of the search is ourselves, the enterprise takes on a dimension absent from other sciences. It is in a sense extra scientific, more philosophical and metaphysical, and it addresses questions that arise from our need to understand the nature of humanity and our place in the world.
Richard Leakey, *Origins Reconsidered*

We are the only animals that dig up the fossilized bones of our ancestors in search of our origins. Where did we come from? It's the million dollar question.

Anthropologists tell us that all traditional cultures rely on a distinct cosmology (a culturally accepted view of how the world came to be) that gives them a sense of identity and an understanding of their role in the world around them. In our modern technological cultures that cohesive viewpoint has been replaced by individual beliefs based on education, religious conviction, and life experiences. Personal beliefs about human origins are especially diverse, and many of us have unanswered questions.

Barbara F.* is a vivacious, well-traveled sixty-year-old widow whose Catholic faith is the bedrock of her life and the inspiration for her philosophy about where we came from.

If you spend a couple of hours looking at those baboons in the zoo, just watching them, you know humans are in there somewhere. You know, evolution.

*Names of interviewees have been changed to protect their privacy.

> *But that's not all there is to it. You can't talk about where we came from without talking about what makes us human — our spirit, our love of life, our soul. And where did that come from?*

Barbara accepts the natural evolution of humans—up to a point. Her strong religious belief prompts her to question how evolution alone could explain the origin of a soul—that one special feature she believes separates us from the monkeys and makes us truly human.

Why do people react so strongly and passionately when asked about human origins?

First, there is the nature of the subject. Discussions about human origins inevitably revolve around some of the most controversial issues in our culture—science, politics, religion, the supernatural, extraterrestrial life, and hope for immortality. Even our unacknowledged beliefs about where we came from can effect the way we live. Are we constantly aware that it's survival of the fittest and dog-eat-dog, or do we believe in a kinder, gentler reason that we inhabit the Earth?

Second, there is the nature of beliefs themselves. Our beliefs are like possessions. We have pieced them together like a patchwork quilt throughout our lifetime. They are the fabric of our lives, the foundation of our virtues and our sins. We guard them, defend them, and sometimes find it painful to share them. They are our most intimate reflections. When our beliefs are challenged, an amazing barrage of mental strategies emerges to help us maintain and protect them—often in the face of overwhelming evidence to the contrary.

Smithsonian anthropologist Paul Michael Taylor explains that the grand cosmology of a primitive culture often gives way to more practical notions of origins. For example, if we asked a member of the Mambai culture of East Timor, "Where do you come from?" they would probably answer, "My people came from the village over the mountain" or "My father's father belonged to the tribe downriver near the fishing place." This local view of origins establishes a social context for the group, a sense of place for the individual, and lays claim to a territory and the political power that goes with it.

I once told a Scottish friend that I was writing a book about where we came from, and he, too, assumed I was writing about the genealogy and immigration of my Scottish ancestors. This local perspective of our origins is very comforting because it doesn't require us to think about the really big questions.

Thus it was with some misgiving that I began asking everyone and anyone I met, "Where do you think we came from?" But in words both awkward and poetic, scores of teachers, teenagers, sales people, scientists, and others shared with me their beliefs and hopes. For some the answers were straightforward and well thought out. Others admitted that they had never been asked the question—but then offered deeply philosophical views hinting at life-changing experiences. These encounters guided my search down new paths and steered me in new directions.

> A young librarian at the Smithsonian Institution who assisted in the early research for this book became curious about the mix of cultural and scientific references I was requesting. I explained to her that I was writing about human origins and asked, "Where do you think we came from?" Her stunning confession of personal belief took me by surprise.
>
> *I always thought I knew the answer to that. I always believed in evolution. But since my father died two months ago, I'm not sure what I believe about it anymore. I feel that wherever my father is now, he knows the answers to all the big questions like human origins, and I wish he could tell me what they are.*

Life-changing events can deprive us of our emotional underpinnings. They change our immediate worldview and cause us to search for understanding and knowledge that might soothe our sense of loss. Our normal mental processes become agitated and we often embrace ideas that previously would have been impossible for us to seriously consider.

Paths to an Answer

National belief surveys consistently report that only 10% of Americans say evolution alone is responsible for life on Earth. Nearly 45% believe in the divine creation of all life.[1]

If someone asked you "Where did we come from? How did life begin?" what would be the first thoughts that crossed your mind? Would it be images of cavemen, or Adam and Eve in the Garden of Eden? Would you envision a divine spark of life similar to the one rendered by Michelangelo on the ceiling of the Sistine Chapel? Or could you picture the deep vastness of interstellar space where star dust forms planets, time begins, and life itself may have first started? What is your view of the world and how we came to be?

All cultures have devised elaborate stories to explain the organization and sophistication of the world around them. Over four thousand years ago at the dawn of their civilization, the Egyptians developed a complex mythology to explain life's most profound questions. How could the Earth, sky, and all living creatures have arisen out of the nothingness that they called the Abyss? An ancient Egyptian coffin text known as "The Book of the Two Ways" promised Egyptians that after they died their soul would have the power to ask these deeply metaphysical and insoluble questions, and that the Universal Lord would answer:

> *I will repeat to you the four good deeds which mine own heart contrived . . .*
> *I made the four winds that every man might breathe thereof . . .*
> *I made the mighty inundation waters that the poor should have rights therein like the powerful . . .*
> *I made every man like his fellow . . .*
> *I made their hearts not to be forgetful of the West . . .*
> *The gods I created from my sweat, but mankind is from the tears of mine eye.*

The Greek story of human origins appears in Plato's Dialogues. In the words of the great philosopher Socrates, the first being was Chaos—"without form and void." Chaos created the gods who in turn assembled the first humans out of earth and fire.

The origin story of the Judeo-Christian religious tradition is strikingly similar:

> *In the beginning God created the heaven and the earth. And the earth was without form, and void.*
> *And the Lord God formed man of the dust of the ground, and breathed into his nostrils the breath of life; and man became a living soul.*
>
> Genesis 1:1-2; 2:7

Over twenty-five hundred years ago, a great awakening of minds arose among a handful of Greek scholars living on the island of Samos in the eastern Aegean. In that stimulating crossroads where Eastern and Western ideas mingled freely, the revolutionary concept emerged that the natural world could be understood without the requirement for a divine or supernatural cause. Modern science was born.

Among those early thinkers was Democritus, who taught that everything was made of *atoms*, a Greek word meaning "unable to be cut." He wrote that life began in primeval mud, and humans and animals developed from simpler forms of life. His fellow scholar Empedocles anticipated our modern concept of adaptation and survival of the fittest by two millennia. He believed that a much greater variety of animals had lived on the Earth in prior ages, but each type had either survived or perished according to its strength and cunning.

The modern theory of evolution, first proposed in 1859 by British naturalist Charles Darwin, remains the most widely accepted scientific explanation for the origin of humans. But Aristotle foreshadowed Darwin when he wrote in his *History of Animals*, "Nature proceeds little by little from things lifeless to animal life in such a way that it is impossible to determine the exact line of demarcation." Aristotle also believed that apes were intermediate in form between humans and other animals.

Darwin's theory that all life on Earth evolved due to the action of natural laws is perhaps the most influential scientific concept of all time. In the one hundred and fifty years since Darwin, science has uncovered much evidence in support of evolution, but questions remain.

Skeptics, True Believers— and a Different View

Many people still satisfy their curiosity about where we came from through belief in the traditional stories of their culture and religion. They embrace the idea that a purposeful and divine act created humans. This creationist view fosters the heady notion that we were made in the image of a god. It has the added benefit of bestowing a unique status on us apart from the other animals. These true believers credit a force beyond nature—a Creator, Supreme Being, or Intelligent Designer—with total and direct responsibility for the origin of the first humans.

In case your own beliefs lead you to assume that by true believers we are referring to a small but vocal minority of religious fundamentalists, consider this. During the past fifty years, Gallup surveys of religious belief in America have consistently reported that over 90% of all Americans believe in God, and over 40% believe that the biblical story of the creation of Adam and Eve is literally true. This latter number jumps to 55% for folks living in the Bible-belt South. Evidence also suggests that a majority of those who don't believe in the literal truth of Genesis nevertheless acknowledge a Supreme Being who guided the origin of humans.

In contrast, modern science exposes us to a much more humbling view of our natural development from simpler forms of life. Biologists tell us that 98% of our genes are identical to those of chimpanzees. Why aren't we still swinging from trees? Are we still evolving?

The question of human origins always leads at some point to mention of religion or spirituality. Atheists are quick to disassociate themselves from any belief in the spiritual. Christian fundamentalists proudly throw down the gauntlet of sacred text and relegate all talk of natural evolution to the realm of the hypothetical.

In his book *Skeptics and True Believers*, Chet Raymo champions the Skeptics as bearers of enlightened truth. Their intellectual leanings put full faith in scientific knowledge. In contrast his True Believers value the security and comfort of a faith-based knowledge of the world. Raymo's categorization of the True Believers includes religious fundamentalists and astrologers, as well as folks who believe in the healing power of prayer or Santa Claus. He argues that the "knowledge" of the True Believers is insupportable and always hypothetical. There is no room for a different view.

Yet the same national polls that measure these two extreme views also report that a whopping 40% of Americans identify with the statement: "Human beings have developed over millions of years from less advanced forms of life, but God guided this process, including the creation of humans."[2]

These are the Moderates. They rise above the *either-or* message of the creationists and evolutionists to allow room for spiritual belief—room for a different view. Why don't we ever hear about this incredibly large segment of our society?

Moderates don't have a spokesperson. They aren't organized, aren't card-carrying atheists, Bible-toting creationists, or zealous evolutionists. Until now they didn't even have a name. Their most striking characteristic is a lack of affiliation with any specific religious sect or denomination.

The voice of this unrecognized multitude is seldom heard above the din of the evolution/creation debate. Meanwhile they comfortably and quietly reconcile belief in the scientific theory of evolution with their personal convictions of faith. Now for the first time you will meet the Moderates and hear in their own words the nature, psychology, and importance of the different view held by millions of Americans.

Today the reactionary rise of "creation science" to counter the influence of the theory of evolution has expanded into a larger social movement whose agenda is to control what we teach our children. Scientists fear this will lead to rejection of objective reasoning. But scientists are part of a broader culture and have their own beliefs and biases as well.

The stunning revelation that what we know and what we believe can be two very different matters came from Don S., a renowned paleontologist at one of America's leading research institutions. After a morning spent discussing the evolution of fish, the bones of early humans, and the rise of creation science in Russia, we had a few moments alone in his lab and I asked, "After all we've talked about today, where do you really think we came from?" There was a long pause, then came the soft reply, "God, I guess."

Oxford professor Richard Dawkins, avowed atheist and defender of Darwinism, has argued against the need for an ultimate cause for human origins in his books *The Blind Watchmaker* and *Climbing Mount Improbable*. On the other hand, physicist and cosmologist Paul Davies has also employed science in his quest to answer the great metaphysical questions, but has come to a somewhat different conclusion. His study of the origins of the Universe and its organizing principles has led him to believe that the Universe and the natural world are not the result of mindless, purposeless forces. In the unanswered questions that lie at the edge of the scientific envelope, Davies experiences a spirituality without belief, a non-religious sense of awe and wonder that leads him to conclude in his book, *The Mind of God: The Scientific Basis for a Rational World*, that "we are truly meant to be here."

The deep chasm between science and religion is underscored by honest attempts to bridge the gap. A recent program on public television convened a panel of scholars, scientists, clergy and laypeople to mend the rift between science and theology. The panelists represented not only the two traditional viewpoints —divine creation and scientific evolution—but also more Moderate positions. Unfortunately, the far more interesting possibilities offered by this third voice in the debate were drowned out by the extreme rhetoric and mutually exclusive positions of the other two. Can there be room for a different view in which science and spirituality coexist? How do Moderates merge the unverifiable truths of their faith in light of the verifiable truths of science?

But wait a minute before you turn off and tune out. You say this mystical stuff is just too much . . . stuff? And you're too logical, too rational to consider the idea of the supernatural, the paranormal, or a higher level of consciousness as part of the answer to human origins? If so, hang on to your seats. One of the hottest areas of scientific study these days is the search for a neurological basis for the soul.

The Nature of Belief

Surely as fascinating as what we believe is how we have arrived at those beliefs. We can sort answers to where we came from into a handful of general categories. These include beliefs based on science, faith in the supernatural, paranormal activities, extraterrestrial origins—and an endless variation on these main themes. I wanted to probe the very personal and often illogical roots of beliefs about human origins. What factors determine how we mold and verify our beliefs? What roles, if any, do family background, religion, education, and life experience play in shaping or changing them? I was not the first to plough this ground.

In 1987 archaeologist Francis Harrold and sociologist Raymond Eve published *Cult Archaeology and Creationism*, a fascinating look at the beliefs of over nine-hundred college students regarding a variety of scientifically unsubstantiated events. Claims put forth for the truth of these occurrences typically rely on unsupported evidence presented in scientific terms. For this reason, belief in their reality is referred to as pseudoscientific belief.

The pseudoscientific beliefs examined by Harrold and Eve fell into two categories: cult archaeology and creationism. Examples of cult archaeology include belief in the lost continent of Atlantis or in Erich von Daniken's theories of earthly visits by ancient astronauts. Creationism or "creation science" refers to belief in the literal truth of the Genesis account of human origins. Harrold and Eve also explored student opinions about paranormal phenomena including psychic powers, UFOs, and communicating with the dead.

Eventually the authors narrowed their focus to beliefs about human origins—specifically evolution versus creationism. They compared the extent of these beliefs among students from three geographic areas (Connecticut, Texas, and California), using several sociocultural factors such as family and religious background, exposure to evolution or creationism in school, and sources the students considered most reliable for scientific information. The result was a unique insight into the role these factors play in development of origin beliefs.

Harrold and Eve concluded that creationist beliefs are positively correlated with what many have termed cultural fundamentalism—a lifestyle and world view that is socially, politically, and religiously conservative.

Perhaps even more tantalizing is what they did not find.

They did not find a positive correlation between belief in creationism and belief in cult archaeology or paranormal phenomena. In other words, creationists were no more likely to believe in cult or paranormal phenomena than evolutionists.

Nor was there a significant difference among creationists and evolutionists in what they considered reliable sources of scientific information. For example, both groups ranked creationist literature very low on a scale of reliable sources of scientific information. Only the *National Enquirer* ranked lower. Furthermore, over 20% of all students held strongly creationist beliefs despite exposure to the theory of evolution in high school or college courses such as anthropology, biology, and geology.

A majority of students believed that evolution has a valid scientific foundation. But 21% to 26% said this foundation is not testable. Furthermore, when asked what the concept of modern evolutionary theory meant to them, one-third of all students (creationists and evolutionists alike) gave answers that indicated a poor or incorrect understanding of evolution.

Other studies reported by Harrold and Eve confirmed that three years of college education, and participation in college level courses designed specifically to teach logical thinking and recognition of hard scientific evidence, did not result in a lasting change in students' creationist belief patterns. It appears that the students arrived at college with their beliefs well formed and strongly internalized. These beliefs persisted despite classroom exposure to evidence to the contrary.

Probing Questions, Personal Disclosures

What is the source of our strongly held beliefs about human origins? How can life experiences alter those beliefs?

In order to delve more deeply into the dynamics that shape beliefs about human origins, I conducted a Belief Survey using questions that fell into six general belief categories. One-third of the questions were uniquely designed to capture Moderate beliefs that merge science and faith. The purpose of the Survey was to focus on a population that chose to participate in the Survey and share their beliefs. It was not used to measure the beliefs of all Americans, but to capture the variety and qualitative nature of the beliefs of the participants.

Survey respondents came from a broad segment of the population aged 16 to 70. They represented a variety of life experiences and included homemakers, students, engineers, and singers among others. The entire Survey Questionnaire is reprinted in the Appendix. The Survey also gave me access to a large number of people from whom I could solicit in-depth, personal interviews about their beliefs. These interviews were semi-structured and conversational. They provided an excellent opportunity to probe for details about the way in which we form beliefs and to differentiate the logical and emotional aspects of our convictions.

Disclosures during the interviews confirmed my theory that most survey findings can only hint at the complex psychology of personal belief. For example, there is evidence from my work and other studies that many folks who respond "yes" to a question that asks whether they believe in "creation" are really saying that they believe in divine purpose—not necessarily in the literal truth of the biblical story of a "Creation." Interviews allowed me to capture these important differences in seemingly similar belief patterns.

Where do *you* think we came from? Find the category in Figure 1.1 that best describes your beliefs about human origins. You can find out if you're in the majority, the minority, or listening to that different drummer by comparing your beliefs with those from various survey results and interviews that appear throughout this book.

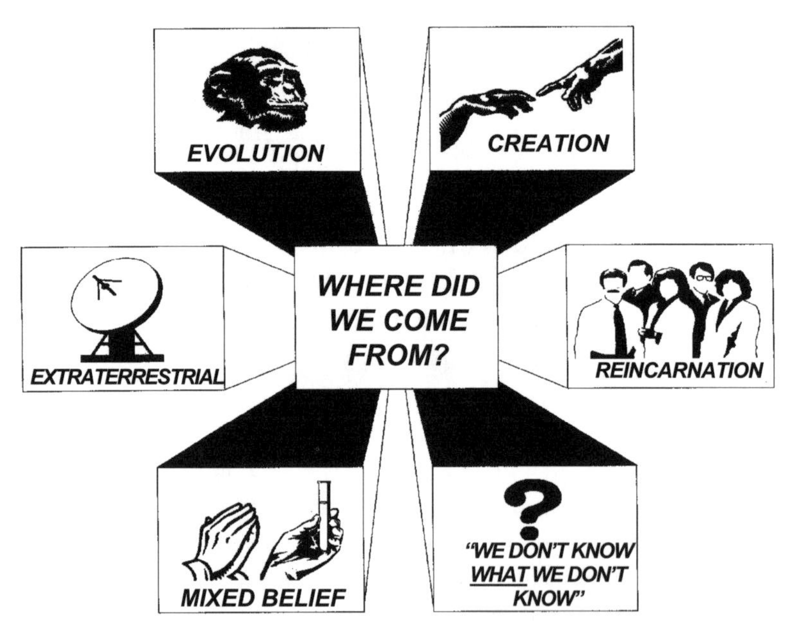

Fig. 1.1 Which category of beliefs would best describe your answer to "Where Did We Come From"?

Today scientists are seeking answers to the spiritual through experiment, and followers of the sacred are opening their minds to science as a path to an enlightened spirituality. The nature of belief is changing.

I once read that the definition of a zealot is someone who having forgotten his purpose simply redoubles his effort. Just for a moment I ask you to suspend all need for a purpose in your beliefs, be it religious, political, scientific, or social, and listen to all the voices in the debate as we travel into the realm of human belief.

The human mind can rationalize anything. But how can people remain strongly committed to contradictory or unproven beliefs about human origins? We'll find out next as we explore the psychology we use to know and believe "what ain't so."

NOTES

1. George Bishop, "The Religious Worldview and American Beliefs About Human Origins," *The Public Perspective*, Vol. 9, No. 5, 1998, pp. 39-44; Gallup polling 1997.
2. Bishop, see above.

REFERENCES

Bishop, George. 1998. "The Religious Worldview and American Beliefs About Human Origins." *The Public Perspective* 9:39-44. University of Connecticut: The Roper Center.

Clark, R. T. Rundle. 1959. *Myth and Symbol in Ancient Egypt.* London: Thames and Hudson. Davies, Paul. 1992. *The Mind of God: The Scientific Basis for a Rational World.* New York: Simon & Schuster.

Harrold, Francis B. and Raymond A. Eve (eds.). 1987. *Cult Archaeology and Creationism: Understanding Pseudoscientific Beliefs About the Past.* Iowa City: University of Iowa Press.

Leakey, Richard and Roger Lewin. 1992. *Origins Reconsidered: In Search of What Makes Us Human.* New York: Doubleday.

Chapter 2

Knowing What Ain't So

Man will occasionally stumble over the truth, but usually manages to pick himself up, walk over or around it, and carry on.
Winston Churchill

One out of every four Americans believe we can communicate with the dead. A comparable number think the sun revolves around the Earth, not vice versa.[1]

Many of us believe strongly in something despite a lack of reliable evidence to support it. Our beliefs are the result of normal mental processes operating on information in unusual ways. We think we base our opinions on facts and experience, but personal bias shapes them. Our emotions conspire with logic to generate beliefs that can persist despite overwhelming evidence to the contrary. Rarely is this more evident than in our beliefs about human origins.

Over three-hundred-fifty years ago Sir Francis Bacon, Chancellor of England and keen observer of human nature, wrote that knowledge has always been influenced by the vagaries of the human intellect. He warned that beliefs reinforced by extraordinary events are unusually hard to dislodge. Our tendency to draw parallels and assume connections where none exist, he argued, ensures that "man always believes more readily that which he prefers."

Before we can weigh the arguments on all sides of the human origins debate, we must take a frank look at our psychology of (mis)belief.

Mind Matters

In his book *How We Know What Isn't So*, psychologist Thomas Gilovich reveals the mental gymnastics we go through to rationalize and defend our beliefs. The title of his book grasps the dual nature of our psychology of belief. That same mental agility we rely on to know right from wrong can turn on us, seducing us into embracing "truth" that just doesn't exist. Four powerful elements in our psychology of (mis)belief illustrate how we unconsciously make mental errors:

1. Misinterpretation of randomness and chance

Random events are independent of all prior events. For example, one hundred random tosses of a coin will statistically yield fifty heads and fifty tails. Each toss has no influence on the outcome of the next. Since a coin has no memory, the chance that a toss of two tails will be followed by heads is still only fifty-fifty. But if we get a long series of tails before heads appears, as shown in Figure 2.1, we sense that something is wrong.

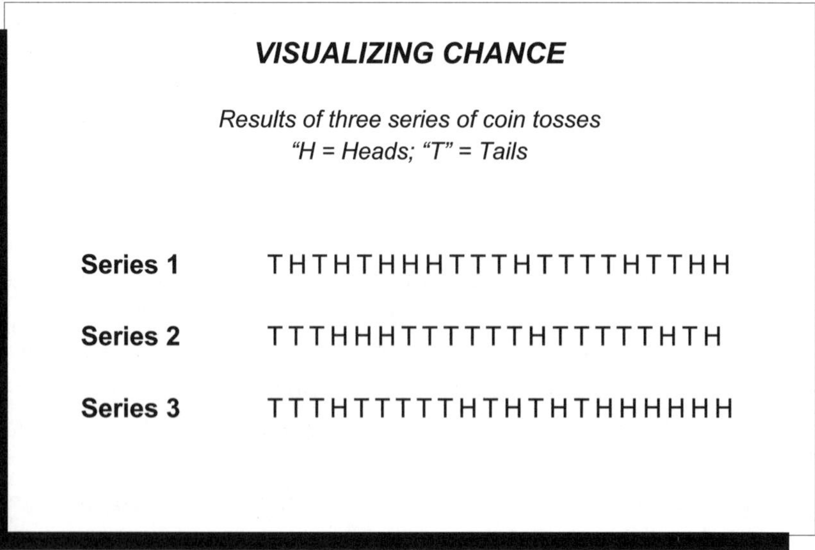

Fig. 2.1 Which of these series of coin tosses looks most random to you?

All three sequences in Fig. 2.1 are randomly generated tosses in which the chance that heads or tails would come up next was always fifty-fifty. Runs of four or five of the same side of the coin are within the limits of statistical probability. But our mind detects a pattern and we believe that if the coin toss is fair and unbiased, then heads (or tails) is due.

The notion that a particular random event is due can lead us to believe that an active cause is governing the outcome of the process. We might say that the cause is explainable (a loaded coin) or unexplainable (bad luck).

It's easy to detect patterns in random events, especially when the sample size is small. We are tempted to associate these patterns with a specific cause. Then we use this cause to explain the outcome of future events. The psychological equivalent of this in sports is the "hot hand."

The phenomena of a "hot hand" is generally accepted by basketball players and fans alike who know the thrill of watching someone who just can't seem to miss. After all, when you're hot your hot! But Gilovich has shown that this widely held belief has no basis in fact. He and his colleagues analyzed National Basketball Association performance statistics, and found no evidence for clusters or streaks of successful shots beyond those determined by chance alone.

Next, Gilovich's group examined the possibility that one successful shot increases the likelihood of success on the next try—the true definition of a "hot hand." They analyzed both the order and outcome of shots at the free throw line where defensive pressure and difficulty of the shot are equal for all players. If success really does breed success, then players making their first free shot should be more likely to score on the second. But the numbers say no. Gilovich found that a player making the first free throw had a 75% chance of making the second, and a player who missed his first try also had a 75% chance of success on the second. Skill is certainly a factor in the success of every throw. But each toss is a statistically independent event, not influenced by the result of the prior throw.

During a college football season match-up between powerhouse rivals Florida State University and the University of Florida, commentators reported that the pre-game strategy of the University of Florida coach was to alternate his two quarterbacks in and out of the game until he saw which quarterback had a "hot hand." Apparently neither had one since Florida State won the game. But belief in the "hot hand" persists, confirming our misunderstanding of the nature of random events and chance.

2. Connecting random events

A kissin' cousin to the "hot hand" is our tendency to actively search for patterns and relationships in random information. Our brain and eyes are adept at drawing connections. It's an important survival instinct that helps us understand and control our environment. We see a similarity among objects or events and assume that like goes with like. We organize random bits of information into coherent patterns, then fashion stories to explain the connections we've made. But this instinct can run amok, leading to errors in judgment and beliefs based on false connections.

In 1894 astronomer Percival Lowell built an observatory in Flagstaff, Arizona to take advantage of the clear viewing made possible by the high altitude and dry desert air. He was eager to observe the "canali" on Mars that had been reported twenty years earlier by the Italian astronomer Schiaparelli.

During the next fifteen years Lowell made countless observations and detailed drawings of the surface of the Red Planet. He became convinced there were canals on Mars and speculated that intelligent beings used them to transport water to their cities from the planet's icy polar regions (see Figure 2.2). Lowell published his theories and maps of Mars in three books—*Mars* (1895), *Mars and Its Canals* (1906), and *Mars As the Abode of Life* (1908).

Over the next several years Lowell made many important contributions to the field of astronomy, but he remains best known for his canals on Mars. Lowell's suggestion that an advanced civilization had built the canals created an immediate sensation and launched our modern day search for extraterrestrial life.

Today we observe Mars through more powerful telescopes and land space probes on the planet's surface. There are no canals,

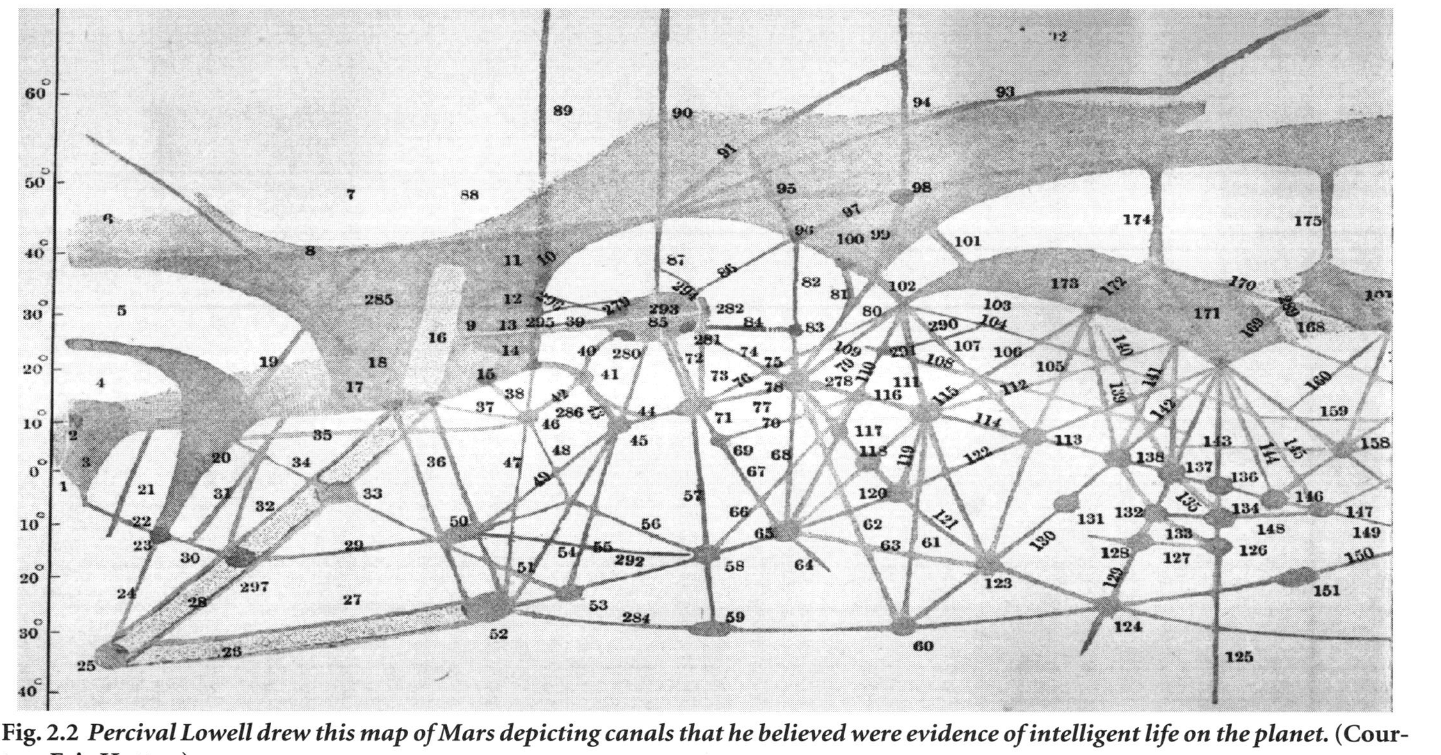

Fig. 2.2 ***Percival Lowell drew this map of Mars depicting canals that he believed were evidence of intelligent life on the planet.*** **(Courtesy Eric Hutton)**

and no evidence exists for intelligent life on Mars. What had Lowell seen?

When Lowell viewed Mars through its turbulent atmosphere, his eyes instinctively drew lines between the darker areas on the planet's surface. The canals were optical illusions, false patterns generated by mental processes. Lowell's theories about Martian cities were stories conjured up to explain the patterns he thought he saw.

3. Biased selection of evidence

Roger M. is a thirty-seven-year-old advertising executive with a degree in computer technology. Roger believes that "someone" created humans, and that the alternative of natural evolution seems improbable.

Objectively and rationally, I have never seen any real evidence for evolution, and the arguments seem to me as much faith based as anything the Bible has to offer.

Since nobody has ever witnessed evolution, real evidence for it will always be circumstantial. Because we are talking about the evolution of life, the fossil record would be the place to look. Now I haven't looked for myself, but I've read a rebuttal to the "missing link" fossil arguments and I've never seen any documentation that provides more than a few bones that have been configured by some paleontologist into a form that would fit the hypothesis of evolution. So I would think there would have to be much more in the fossil record to support a comprehensive theory of the evolution of all species that exist today.

It's been said that *bias* is when the opinion of the *other* person is different from our own. Roger's arguments illustrate an extreme form of bias captured in song by the 1940's singing sensation Bing Crosby when he crooned Johnny Mercer's tune, "You gotta Accentuate the Positive and Eliminate the Negative."

Roger takes the Bing Crosby approach to the evidence for human origins. He accentuates any argument in favor of his belief in creation—and eliminates the negative. While interviewing Roger,

I got the impression that there could never be enough of the right kind of evidence in support of evolution to change his bias against it. His will to believe in divine creation is so strong that he sifts through facts and arguments to find support for what he believes, then eliminates the rest.

Human beings are innately self-serving. Our bias of self interest plays a major role in forming and protecting our beliefs. We think that our opinions are more objective and less emotional than others. We select books and articles that support our beliefs. We socialize with people who share our values, and share our opinions with those hold the same point of view. As a result we are exposed to a biased sample of arguments and evidence that builds a false consensus for our beliefs.

Einstein once referred to common sense knowledge as the collection of prejudices acquired by age eighteen. Psychologists tell us that we can only hold eight or nine pieces of information in working memory at any given time. From these small samples we draw sweeping conclusions that we refer to as common sense. Since we rarely have access to all the information, many of these generalizations are wrong. Common sense knowledge is often uncommon sense because it doesn't accurately reflect the bigger picture.

4. Rationalizing negative evidence

What happens when we're confronted with undeniable evidence that our beliefs are wrong? Darwin thought that most people simply ignore negative evidence. But studies have shown that our most natural response is to closely scrutinize this kind of evidence in order to explain it away. Then we refocus the argument on information that supports our claim.

This recipe for rationalization was used recently by a friend to substantiate his belief in the "hot hand." First he argued that Gilovich's negative findings from professional basketball were interesting but limited, and probably were not generally applicable. Then he countered with evidence for existence of the "hot hand" in other sports, citing streaks of base hits and home runs by celebrated baseball players. Finally he explained that the cause of the "hot hand" is known to be "success through confidence building," a phenomenon which he has personally witnessed during thirty years in business.

Despite the objectivity of modern computer analysis, and the much heralded use of the scientific method, our mental processes and personal motivations still get us in trouble. Faulty reasoning and biased assumptions go hand in hand. Professionals explore these dynamics of belief, but admit that it's often difficult to separate psychological from sociological components. The human mind—that exulted source of pure logic and objectivity—can rationalize anything it wants to believe.

The Nightly News

A popular joke among scientists claims that 97% of all statistics are made up. We say that figures lie and liars figure, but numbers and statistics are respected more often than suspected. The media uses them to its advantage in the *factoid*—a snippet of statistical information that startles us into belief. Factoids range from the silly to the serious. For example:

- Hot dogs are among America's favorite foods. Each American consumes on average 60 hot dogs a year! (*National Hot Dog & Sausage Council*)
- Global fish production exceeds that of cattle, sheep, poultry or eggs, and is the biggest source of protein in the world. (*Ocean Voice International*)
- There are more than 1,000 chemicals in a cup of coffee. Of these, only 26 have been tested, and half caused cancer in rats. (*Dr. Bruce Ames, UC Berkeley*)
- During the first year of background checking, 160,000 people were prevented from buying guns. Three out of four were convicted felons. (*Department of Justice*)

Factoids carry the ring of truth because their message is unambiguous. Their statistics aren't usually questioned if widely quoted and attributed to reputable sources. But buyer beware. Consider this tidbit:

- According to the FBI, a woman is beaten every 15 seconds.

This surprising statistic has been used to stir debate and prompt legislative action. The truth is that the FBI doesn't collect or track statistics on domestic violence. Though the FBI did estimate that a woman is beaten every 15 seconds, it obtained that number from studies reported in *Behind Closed Doors* by Murray Straus, Richard Gelles, and Suzanne K. Steinmetz. Other sources list various numbers of seconds depending on how a particular study was conducted, what constituted abuse or violence, and whether a study was done at all.

We are also more likely to accept at face value what we hear on the nightly news, from the FBI, or from our local doctor than from our next door neighbor. Secondhand information, in general, has a strange influence on our beliefs. It's as though the truth of what we hear is validated by the retelling.

Just the Facts, Please

Numbers and statistics can paint powerful images. But they can also be very misleading.

For example, a study that followed the arrest histories of 4,000 men born in Copenhagen between 1959 and 1961 found a direct correlation between mothers who smoked and the long term criminal behavior of their sons. Men born to mothers who smoked more than 20 cigarettes a day during their last three months of pregnancy were twice as likely to be arrested for violent crime and become lifelong offenders.

As in all correlation studies like this one, there were three possible conclusions:

1. **A causes B**

   Smoking damages the central nervous system of the fetus in utero, leading to deviant behavior (the conclusion).

2. **B causes A**

   Violent adult males caused their mothers to smoke while pregnant (impossible).

3. **C (a third unknown factor) causes both.**

Researchers controlled for variables such as socioeconomic status, parents' age and psychiatric profile, and the father's criminal history. But many unknown factors may have influenced the results, such as the prevailing morality of the home environment or the mother's treatment of her children.

Correlation studies are popular for the same reason that they can be misleading. The exact cause of the correlation often remains vague. Depending on the quality of the study there can be many factors that interact to produce the result. It's easy for our nimble minds to find a statistical link between two independent pieces of data (smoking mothers and criminal sons) and develop a good story that explains cause and effect.

Where There's a Will

"If you think it, you can do it" is a trendy slogan seen on T-shirts and talk shows. But this alleged power of our will becomes physical reality in the placebo effect. Placebos are sugar pills or dummy doses of medicine used in clinical trials to test the efficiency of new drugs. Patients in the trial don't know if they are taking the drug or the placebo. For many years doctors have known that 5% to 10% of all patients given a placebo will show clinical improvement similar to those receiving the actual drug. This effect can be measured, and most drug studies assume some placebo effect and establish controls for it. But in certain instances the effect can be unusually high. For example, Viagra pills, the breakthrough drug for treatment of impotency, have been shown to have a placebo effect as high as 25%. Talk about mind over matter!

Not only can the placebo effect be quantified, but now researchers have begun to unravel how it works. Brain scans done on volunteers who have taken either a narcotic pain-killer or a placebo show heightened activity in the same regions of the brain. These areas are abundant in receptors for pain killers, and also include portions of the cortex known to process emotions. Not only does this demonstrate a physical association between the action of painkillers and placebos, but also leads researchers to conclude that the emotional and physiological components of pain killing are deeply intertwined.

The placebo effect is a self-fulfilling prophecy that requires little action on our part. We don't consciously change our physiology. The unconscious mind does it all. A version of the placebo effect may operate to reinforce our beliefs when there is little evidence to support them. If we see a shred of positive evidence, then our belief is disproportionately confirmed. The weight of negative evidence diminishes. We take the sugar pill and get well. Who can argue with that?

The Madness of Crowds

There is an irresistible power of persuasion embodied in the actions and convictions of our peers. Even the most independent thinker is sometimes reluctant to be considered different from the crowd.

British historian Charles Mackay first described the undue power of crowd psychology in his 1841 classic *Memoirs of Extraordinary Popular Delusions and the Madness of Crowds.* Mackay's tales of human folly include the fraudulent land investment schemes of the 18th century gambler John Law, the financial debacle known as the "South Sea Bubble," belief in the alchemist's ability to turn lead into gold, and the burning of witches. Perhaps the most frightening of all is Mackay's account of popular avarice and group contagion embodied in the "tulip mania" of Holland in the 1600s.

Tulips were first introduced into Western Europe from Turkey around 1559. In fact, the word "tulip" derives from the Islamic word for turban. As tulips became highly prized for their rarity and beauty, the demand and price of bulbs skyrocketed. Collectors vied to own the rarest for their gardens. Some speculators spent up to half their personal fortune to obtain a single bulb. Many jumped on the bandwagon as the market value of tulips rose. Aristocrats, merchants, chimney sweeps, and farmers sold their houses and lands and invested in bulbs.

But at last prudent minds prevailed. A few of the wealthy stopped buying and hoarding the bulbs, and started selling them for profit. Sellers soon outnumbered buyers, and the tulip bubble burst. Investors lost their shirts as tulip mania turned to panic.

This scenario of four centuries ago resembles the up and downs in the stock market today. Investors pour money into high tech companies long before they show a glimmer of profit. The convictions and momentum of the crowd wins out over sound reasoning.

The power of group thinking and peer pressure can also build a false consensus for common misbeliefs about human origins. We garner social support for views that may have little basis in fact. Like tulip mania, our beliefs take on a life of their own.

Believing is Seeing

Deception isn't the sole province of liars, cheats, and criminals. It can become the agent of innocent people wishing to turn belief into reality.

At the dawn of the 20th century no fossil remains of early humans had been found in Africa. Bits and pieces of Neandertal skeletons had surfaced in Germany and France in the mid 1800s, and Rene Dubois had uncovered Neandertal-like "Java Man" in Asia in 1891. Where was the English Neandertal?

In 1912 Charles Dawson, a lawyer and amateur paleontologist living at Piltdown in Sussex, England, found several fragments of a fossil skull in a gravel pit near his home. The deeply stained bones had the rounded contours of a modern human skull but appeared to be very old. Dawson took the find to Arthur Smith Woodward at the British Museum and asked for a professional opinion on the age and identity of the bones.

Woodward reported that the skull appeared to pre-date the French Neandertals, which he gleefully pointed out were probably a degenerate offshoot of early humans. He accompanied Dawson to the Piltdown quarry, and together they unearthed more fragments, including an ape-like jaw with molars that were worn flat like human teeth.

Dawson and Woodward named their discovery "Dawn Man" (*Eoanthropus*) and hailed it as the missing link between apes and humans. Modern humans had indeed arisen from more primitive stock, they said, of which Dawn Man was the evidence. Humanity had taken its first steps in merry old England.

Woodward unveiled his recreated skull of Dawn Man to a generally receptive crowd at the Geological Society in London. But there were skeptics from the start. No one doubted the skull's age, but some suggested that the bones of two different animals had been accidentally mixed together in the quarry.

During the next three years additional fragments turned up that seemed to confirm the authenticity of the Piltdown find. Finally in 1915 Dawson appeased remaining skeptics by unearthing more pieces of thick skull bone, and another apish tooth, at a location several miles distant from the original find. The evidence was mounting, and Dawson's detractors soon became believers. When Dawson died unexpectedly in 1916 at the height of his celebrity, many prominent scientists had already bet their careers on his find. Dawn Man kept a firm grip on anthropology for another thirty years.

Then in 1949 British anthropologist Kenneth Oakley, using a new fluorine method to date the Piltdown find, proved that the bones were no older than 600 years. Oakley also reported that the cranium was human, but the jaw was from an orangutan. At first he didn't suspect a fraud, and presumed that young bones had been accidentally buried together in ancient soil. But by 1952 Oakley had become convinced that Piltdown was a sophisticated hoax (see Figure 2.3).

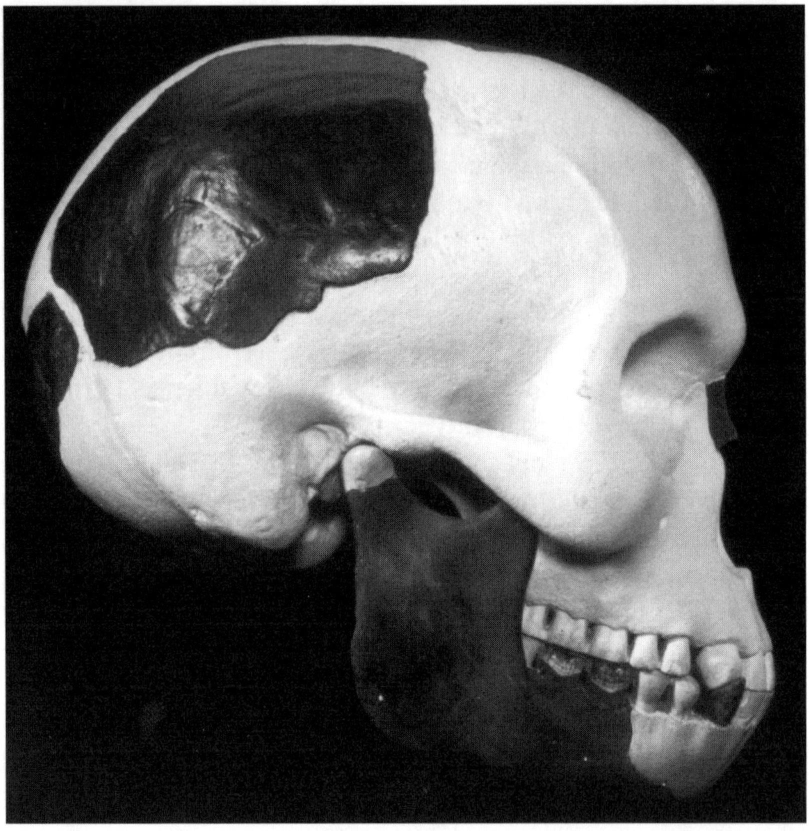

Fig. 2.3 Reconstruction of the Piltdown skull. Fossil bones are seen as dark areas. The experts said the find was genuine, but should they have known better? (Copyright The Natural History Museum, London)

Further analysis revealed that the Piltdown bones had been artificially stained with iron and manganese. The teeth, flint blades, and bone implements found alongside the bones had been filed down using modern tools. The orangutan jaw had been carefully broken to eliminate the chin and the point of articulation with the skull—key pieces of evidence that would have proven that the bones could not have come from a single source.

Dawson, Woodward, and a list of suspects that read like a Who's Who of British science were accused of perpetrating the fraud. The identity of the culprit remained a mystery until 1996 when a cache of old bones were found in a trunk in the basement of the Museum. They had been stained with the same chemicals as the Piltdown find. The initials on the trunk were M.A.C.H.

Martin A. C. Hinton had been a talented curator in the museum's zoology department. Rumor had it he carried a grudge against Woodward over a pay dispute. Could Piltdown have been Hinton's act of revenge? Or was it simply a practical joke that got out of hand?

Piltdown was no laughing matter. The hoax stymied the search for human origins for decades, wasting scientists' time and energy. Once exposed, Piltdown appeared to be crudely forged and embarrassingly obvious. Why had it been so successful?

Like most famous frauds, Piltdown's success had been guaranteed by the psychology of (mis)belief. Biased selection of evidence and the power of peer pressure conspired to interpret the facts and evidence to fit what everyone wanted to believe.

For example, in an amazing feat of rationalization, the anomalous ape jaw had been reconciled with the human cranium by pointing out ape-like features in the human skull. This approach not only endorsed Piltdown as the "missing link," but also played into the belief that humanity's roots would be found on English soil.

The Bing Crosby approach ensured that positive evidence would be emphasized. Negative evidence was disregarded from the beginning. When the first canine tooth was found, a dental anatomist confidently stated that it was too young and showed too little wear to belong to the same jaw that held mature, well-worn molars. His comments were ignored.

But the principal reason Piltdown endured for so long was that professional access to the evidence was restricted. Casts were available for study, but they would not have revealed the artificial staining and shaping of the original bones. After the hoax was exposed and the evidence made public, scientists immediately pointed out the obvious chisel marks on the teeth.

Unfortunately we can't banish scientific forgeries to the past. In the late 1990s an extraordinary 124-million-year-old fossil resembling a small lizard with feathers was smuggled out of China and into the hands of excited paleontologists. Many hailed it as the missing link between birds and dinosaurs. But in early 2001 an x-ray of the specimen revealed that 88 pieces of rock containing the bones of five different animals had been glued together like a jigsaw puzzle to create a faux fossil. This Piltdown Turkey turned out to be a primitive bird with a meat-eating dinosaur's tail.

When Science Replaces Sensationalism

Let's once and for all dispel the notion that science is immune to the psychological pitfalls of misbelief. Science is simply a way of understanding and explaining the world around us. It is a process that the Nobel prize winning chemist Linus Pauling called "the search for Truth."

Among the first Greek scientists twenty-five hundred years ago, any explanation that sounded logical and stood the test of common sense was considered valid. The more creative the better. For example, Thales of Miletus agreed with ancient theories that the Earth had once been completely covered with water. But Thales said that the appearance of dry land wasn't divine intervention but a natural process—a build-up of soil similar to the silting effect he had seen in the Nile delta. Thales' conclusion was wrong, but his method of reaching it was sound. He had found a solution based on observation of a natural phenomenon, not faith in a divine act.

But for sheer volume of information and unbounded imagination nothing surpassed the runaway best seller *Natural History*, compiled in 77 A.D. by the Roman poet Pliny the Elder. Pliny's thirty-seven book encyclopedia reads like *The Farmer's Almanac*. He managed to pull together most of the known wisdom on everything from astronomy to zoology, and growing onions to curing dandruff. *Natural History* remained the undisputed authority on all things great and small well into the Renaissance. In Pliny's writings, fantasy and reality coexist. For example, he wrote that startled porcupines shoot their quills (they don't), that the world is inhabited by dog-headed people that bark (it isn't), and that "absent people can divine by the ringing in their ears that they are the object of talk" (a superstition that lingers today). Pliny eventually lost credibility in the late 15th century when navigators opened up the world to scientific exploration.

The 17th century philosopher Francis Bacon is credited with originating the process of empirical observation that defines modern science. But his emphasis on amassing copious amounts of evidence without a preconceived theory was an intentional exaggeration, designed to stress the importance of factual information over biblical authority.

Charles Darwin claimed that he had adopted Baconian principles when he began to study the origin and diversity of life. But later writings clearly reveal that Darwin, like all scientists, had some idea of what to look for, and how to look for it, before starting his research.

Today scientists incorporate theory up front, then seek proof or disproof through observation and experimentation. Theories go in and out of fashion as new evidence is uncovered. Eventually a single theory triumphs when the weight of the evidence confirms it. This process of acceptance is rigorous and prolonged at best, but for truly innovative theories it can take a lifetime.

In the early 1900s Alfred Wegener thought he had found the answer to a problem that had plagued biologists and geologists for decades. Fossils of hundreds of closely related animals and plants had been found on continents and islands separated by impassable mountains and deep oceans. Coal deposits formed by ancient tropical forests had been discovered in Antarctica, and the telltale signs of the movement of glaciers were still visible in the Sahara. How had they gotten there?

These marvels implied drastic changes in climate and geography during our planet's history. Most attempts to explain them started with the premise that continents don't move. The puzzling distribution of animals and plants was attributed to ancient land bridges, a convenient catchall solution not easy to discredit.

Alfred Wegener had a more radical idea. What if the continents themselves had moved? Wegener suggested that millions of years earlier the continents had been arranged over the Earth's surface in a very different pattern. As the continents moved they had collided and merged into a single giant landmass that later broke up and scattered around the globe. This action alone would account for signs of ancient glaciers in desert regions, and the distribution of similar plants and animals across lands separated by thousands of miles of water.

Wegener was immediately branded a crackpot and his ideas dismissed as impossible. But he persisted, and by the 1930s had fleshed out his theory of continental drift. All he needed was a plausible mechanism. His own hypothesis was that gravitational forces of the sun and moon had put the continents in motion, dragging them east to west as the Earth turned. But this scheme was quickly rebutted by physicists who calculated that the forces involved were far too weak. One of Wegener's supporters suggested a mechanism involving radioactive melting of continental edges, resulting in a lubricated skidding of the landmasses on top of the planet's crust. Ideas like that one made Wegener's theory look even more ridiculous.

It wasn't until the late 1960s, with Alfred Wegener long dead, that technology finally caught up with theory. Advanced submersible instruments drilled cores from the ocean floor, and geologists finally confirmed that volcanic forces are still molding the Earth's crust and moving continents. Hot magma swells up from deep inside our planet through cracks in the ocean floor and separates the crust into plates. As the magma cools it forms new ocean floor and pushes the old crust away from the fault line inch by inch. The continents, resting on the moving plates, sail slowly through the oceans like ponderous ships. This concept of plate tectonics (literally "plate building") validated Wegener's theory. It easily explained the old evidence of fossil distribution and climate change. Wegener's theory had triumphed.

The embryologist Karl Ernst von Baer once suggested that innovative scientific theories like Wegener's must pass through three stages as they become accepted. First they are discarded as wrong or impossible, then ridiculed as profane, and finally embraced as dogma, especially by former critics who now claim they had seen the possibility all along. These dynamics of the scientific process frustrate our psychology of belief because:

- *Science rarely reaches ultimate conclusions or commits to final answers.*

  The verbs of science are rethink and retest.

- *Science advances more by disproof than by proof.*

  The last theory left standing wins. Science embraces debate and offers few certainties. This self-correcting process moves the science of evolution forward, and stands in sharp contrast to the certainty without proof of the "creation science" movement we'll expose later.

- *Science is a piecemeal process—question, theory, experiment, result, conclusion.*

  Many scientific questions seem simple and straightforward. In fact, most are extremely complex. They require multiple theories and diverse approaches to solve the problem. Initial findings often result in conflicting conclusions, inconsistent data, and confusing press releases. Science tests our patience, and too often pseudoscience rushes in to meet our needs.

Numerous surveys have revealed the public's lack of scientific knowledge and inability to comprehend the scientific process. In 1996 the National Science Board asked a representative sample of adult Americans about their attitude toward science and their grasp of basic scientific information. It found that only 47% of us know that the Earth revolves around the sun once a year, and only

48% believe that dinosaurs and early humans lived at different times. We score better on other questions (91% of us know smoking causes lung cancer) but have a lot of trouble explaining how science works. Only 2% of those surveyed said they believe that science is the development and testing of theories. Another 34% link science in some way to experiment, comparison, or measurement. It's not clear what the remaining 64% think scientists do. Against this vague understanding of science, it's not surprising that many folks can't differentiate between the science of astronomy and the pseudoscience of astrology.

Over 80% of those who responded to my Belief Survey claim they have a clear understanding of the scientific process—not surprising when we consider the sizable number of college graduates (48%) and science or engineering majors (60%) among participants. Overall attitude toward science is also positive—70% agree that science has done more good than bad for the world. More reassuring is that over 76% say they don't believe that astrology can predict the future. It would be interesting to know how many in this group read their daily horoscope.

What is the agenda of science? *Science is in the business of fact, not faith.* Does this mean that science invariably leads to a clear and objective solution as it did for Alfred Wegener? Eventually, yes. But consider the case of salt.

Table salt, or sodium chloride, is a requirement for smooth running of our neuromuscular and circulatory systems. It is essential for life. Salt is also a hallmark of the advancement of civilization from nomadic culture to the settled lifestyle of an agricultural society. The diet of nomadic herders consists mainly of milk and roasted meat and needs no salt supplement. But farmers eat mostly vegetables, grain, and boiled meats whose salt content has been lost. They must add salt to their food for the same reason that cattle lick salt to supplement their diet of grass.

Salt was highly prized in ancient times. Cakes of salt were used as money in Africa and Tibet for centuries. Roman soldiers were given an allowance of salt or the money to buy salt that they called *salarium*—"salary" in English.

Salt was used as a preservative and soon became symbolic of endurance and faithfulness. It was an important part of sacrificial meals and the sealing of contracts. The word for salt became synonymous with honor and esteem, a connotation that survives when we say someone is the "salt of the Earth." But in today's health conscious culture, salt is a four-letter word. For the past three decades we have been warned that eating foods rich in salt causes high blood pressure.

The controversy over salt began in the 1970s when public health organizations began focusing on the devastating effects of high blood pressure and the heart and vascular diseases associated with it. Early studies had shown a direct link between high salt intake and high blood pressure.

It's a physiological fact that when you eat an unusually high amount of salt, your body retains both the salt and a proportionately high volume of water to dilute it. The additional volume of fluid causes a slight rise in blood pressure in most people. Eventually the kidneys excrete the unwanted sodium along with the water. Not all people respond this way, and some people appear to be more sensitive to the salt effect than others.

But many factors besides sodium are involved in the complex mechanism that maintains reasonable blood pressure. For example, high calorie diets can also squeeze the blood vessels and raise blood pressure. If you have ever gone for a long period of time without eating and feel faint or dizzy, you have experienced the opposite response—not enough calories lowers blood pressure. Furthermore, the way these systems interact and the extent of their response varies with age, sex, genetics, and race. Nevertheless, most doctors believe that high salt diets will lead to higher blood pressure and salt restricted diets will encourage lower blood pressure.

So what's the problem? Well, for starters, most early evidence for a direct association between salt intake and high blood pressure was sketchy or anecdotal. It wasn't until the seventies after the dietary taboo against salt had been established in public health lore that serious studies were conducted to prove the link. After a decade of research the evidence remained inconclusive. Some results had done more to disprove the connection than support it. Finally in the 1980s two important investigations were launched to resolve the issue once and for all.

In 1988 the Scottish Heart Health Study reported that sodium had little or no effect overall on blood pressure among 7300 Scottish men in the survey. Advocates of salt restriction promptly dismissed the troublesome findings by explaining that the sample size (7300 people!) was too small to overcome problems seen in similar experiments. A follow-up study ten years later by the same Scottish team again showed no relationship between salt intake and heart disease or death.

The most influential and controversial salt study, involving years of effort by 150 investigators studying over 10,000 men and women in fifty-two communities around the world, failed to confirm a direct relationship between high salt intake and high blood pressure. In 1988 this study reported it had found a slight indication that reducing salt intake would cause a slight reduction in the normal rise of blood pressure with age. Though most investigators thought the study's findings didn't support salt-restricted diets, the anti-salt forces rallied successfully once again.

Today the interpretation of evidence for or against salt restricted diets seems to depend largely on the beliefs of the investigators. The Bing Crosby approach succeeds because results from salt-versus-blood-pressure studies haven't been consistent or conclusive. Anti-salt forces rally at the slightest evidence that favors a link between salt and high blood pressure. Additional studies that fail to find support for this view get little press. Tension remains high between pro-salt commercial interests and government health programs that aim for the potentially large benefit that salt restriction could have on public health.

The salt debate underscores the message that science is a human endeavor and prone to the fallibility of human emotions. A strong will to believe inspires creative interpretation of sketchy evidence. One glance at the "Letters" section of any prominent scientific journal will expose the ways in which scientists sometimes react to unwanted results. Dismissal of negative evidence by blaming experimental error or a hasty retreat to entrenched theories are not uncommon. All extraordinary scientific truths must survive this gauntlet.

Science is also about cause and effect. Scientists delight in finding patterns in information. But anticipation of these patterns in experimental results can lead to biases known as *expectancy effects* —the scientific equivalent of self-fulfilling prophecies.

The design of any experiment anticipates a certain outcome that will prove (or disprove) a theory. Assumptions about the outcome may inadvertently influence how investigators interpret the results. These biases are insidious and particularly difficult to avoid in experiments involving humans.

One of the best ways to protect experimental results from expectancy effects is to use double-blind methods—a routine practice in clinical trials of new drugs. Double-blind procedures provide protection by shielding both the doctor and the patient from knowledge about whether the drug or a placebo is being administered.

A good imagination is essential to a scientist who must be able to grasp the significance of an initial observation, develop a theory to explain it, then test that theory. But science relies on rules and procedures that reign in the freewheeling human intellect and protect science from bias. All results must be replicated by other scientists, in other laboratories, and in other settings. One experimental result is not sufficient for scientific proof. These rules of the scientific process can pose a problem for the study of evolution. It will never be possible to rerun four billion years of life on Earth. But innovative technologies are testing the mechanisms of evolution, and confirming our theories about the development of life on Earth.

Science is still the best game in town. In any age the scientific process, as openly rancorous as it can get, remains the best route to knowledge. The 18th century chemists believed that heat is a fluid. The 19th century geologists insisted that continents never move. Early in the 20th century, biologists claimed that genes are made of protein not DNA. We now know that these scientists were wrong in their conclusions, but not unjustified in making them. Their interpretations were scientifically reasonable. They were based on the best available evidence and technology. As Robert Pirsig wrote in *Zen and the Art of Motorcycle Maintenance*, "the real purpose of [the] scientific method is to make sure Nature hasn't misled you into thinking you know something you don't actually know."

The Power of Knowing What Ain't So

> *Finding the occasional straw of truth awash in a great ocean of confusion and bamboozle requires intelligence, vigilance, dedication and courage. But if we don't practice these tough habits of thought, we cannot hope to solve the truly serious problems that face us . . . and we risk becoming a nation of suckers, up for grabs by the next charlatan who comes along.*
>
> Carl Sagan, *The Fine Art of Baloney Detection*

Skepticism is not in the job description for most of us. Being an incurable skeptic would be out of character. But for some people, skepticism is a life's work.

James Randi is a magician. Performing as "The Amazing Randi," he has thrilled audiences around the world with his sleight of hand and daring escapes. But for over forty-five years Randi has been on a personal mission to investigate and debunk claims of the supernatural and paranormal. His knowledge and skill in the art of illusion has made him a confirmed skeptic of UFOs, ESP, levitation, psychic surgery, and other phenomena akin to magic and miracles.

In many cases, Randi has been able to convincingly recreate alleged paranormal or supernatural phenomena simply by using the tricks of his trade. For example, he was able to debunk the psychokinetic feats of Uri Geller, an international celebrity famous during the 1970s for bending spoons with the power of his mind. Randi replicated Geller's performance using simple sleight of hand. When Randi tested Geller's psychokinetic abilities under conditions in which the "trick" could not be done, Geller's performance failed.

The popularity of beliefs in the supernatural and paranormal lies in the nature of evidence. The average person is often unable to distinguish between solid evidence from good science, and something that resembles science but is based on anecdotal evidence and uncontrolled experiments. If false information is frequently repeated, particularly in the guise of a "documentary" program, we believe there must be a kernel of truth in it. It's all "evidence" to us. If it sounds like science, all the better. There's a big dose of the will to believe at work here that overcomes our skepticism. Who wouldn't want to possess the untapped potential of psychic power, or be distracted from the routine of daily life by visitors from outer space?

James Randi destroys the illusion of the supernatural and paranormal by making the steps clear to us. We are entertained by what we see, but we aren't fooled into thinking that we are witnessing the miraculous or unexplainable. This is also the role of science.

A healthy dose of skepticism is science's self-correcting mechanism. Good scientists must be practicing skeptics. Are you a skeptic by nature? On a scale of 1 to 10, how alert is your belief radar when it comes to questioning the evidence? Let's look at seven powerful tools we can use to boost our reasoning skills and avoid "becoming a nation of suckers."

Seven Habits of a Healthy Skeptic

1. *Consider the source.*

Bold or outlandish statements do not make a claim true. Who or what is the source of the claim? Is a well known expert giving an opinion based on solid evidence and a proven track record? Or are we hearing a fascinating but virtually fact-free theory about the survivors of the fabled city of Atlantis?

This latter scenario is often the most dangerous. Cunningly crafted by a sensationalist provocateur and emceed by a credible celebrity, the information is presented in docudrama tones designed to persuade. The tale seems genuine unless you are familiar with the historical, cultural, or scientific background and are able to separate fact from fiction. Most of us aren't, so we listen and sometimes believe. When outlandish claims go unchallenged by professionals, and media coverage lends authenticity, more than a hint of credibility creeps in. This has been the case with many claims by "creation science" against evolution.

2. *Contrary opinions aren't necessarily right, just different.*

Physicist Niels Bohr, winner of the Nobel Prize for unraveling the structure of atoms, is said to have commented to a graduate student during his doctoral defense, "We all agree your theory is crazy, but is it crazy enough to be believed?"

Not all crazy ideas have merit. In fact, most don't and their claimants fail to show proof time and again. But there is a tendency to rationalize those failures by saying that all great ideas were once ridiculed. That's often the case. But those same great leaps of knowledge have also met the burden of proof. If we forget the times when claimants failed to read minds or bend spoons in controlled environments, and only remember the incidents promoted as successes, we have been bamboozled.

3. *Consider the evidence.*

Quality as well as quantity of evidence supports the burden of proof. What type of information supports the claim? Does it come from unreliable eyewitness accounts or controlled experiments? Does it consist of a few initial results from a small study or the statistically valid conclusions of several studies. Or is it simply hearsay?

4. *Absence of evidence is not evidence.*

If the evidence doesn't support one theory, it doesn't prove the opposite. This notion of "either/or" presumes that the truth is limited to two prevailing ideas. The experimental technology needed to prove a theory may be decades in the future, as it was for

Alfred Wegener. Absence of evidence is not evidence. Nor is it evidence of absence, as we'll see when we consider the possibility of life on other planets.

5. ***Does the claim invite challenge?***

The scientific method, a rigorous process designed to defy deception, insists that experimental results be replicated. When claims are made for a breakthrough discovery, look for signs that its advocates are encouraging others to repeat their experiments or examine their evidence. Are they publishing their research in a timely fashion in peer-reviewed publications? Beware if their methods are cloaked in secrecy rather than laid bare for all to challenge. Remember the excitement over the discovery of "cold fusion" in the 1980s? If you don't, it's because it never really happened.

6. ***Look for smoking-gun evidence.***

Perry Mason was right. The case is always easier to win if you have the smoking gun as evidence. Unfortunately a lot of what we know today is based on complex mathematics and arcane reasoning that would foil the most cunning lawyer. What kind of smoking-gun evidence should we expect to see that would confirm or deny the authenticity of a claim?

In every case ask yourself, "What one piece of information could I reasonably expect to hear or see that would convince me that Atlantis did exist?" There might be three or four things that would add to the weight of positive evidence in your mind. Look for these in announcements. Listen for these in conversations with friends on any subject when your skeptic's antenna goes up.

7. ***The human mind can—and will—rationalize anything!***

Some questions don't lend themselves as neatly to the rules of scientific inquiry as others. "Where did we come from?" is that type of question. Problems that demand extraordinary evidence, countless variables, or intricate connections can be just plain messy. Solving them requires drawing difficult inferences and using that finely honed skill called rationalization.

Our ability to rationalize also enables us to take almost any given piece of evidence or information and use it to support what we want to believe. In the political arena this maneuver is called "spin." In the arena of knowing what ain't so, it's called poor judgment. Spin bends the rules of logic and encourages our shifty psychology of belief to go where no truly rational mind would go. Often it's hard to recognize where good judgment stops and spin begins. Most of us fall victim to spin because it feels good to have an answer we can accept.

But to be an honest skeptic you must face the possibility that at times your mind is spinning your beliefs. It's nearly impossible to defend against mind spin, but you can begin by asking yourself:

- If the results had been the opposite, would I have used that information to support my belief?
- If I were asked to debate this question from the other side, how would I use the same evidence to win the argument?
- Is there another solution, another theory different from those we are arguing, that could satisfactorily explain all the evidence?

Granted, it can be hard to be a skeptic. But if you try it, you'll have a much deeper insight into the power of your own psychology of belief and knowing what ain't so.

Where do you think we came from? Just turn the page and take the summary version of my Belief Survey (Figure 2.4).

How did you do on the summary Survey? Were there some questions that you had never thought about before? Were your answers swift and sure, or did you have to think long and hard about your true beliefs? Were you skeptical about what you believe? Do you think a lot of other people would give the same answers as you did, or do you feel that your views are unique?

Next we'll take a look at the psychology of belief at its best (or worst) as we examine the controversies surrounding evolution.

BELIEF SURVEY (Summary)

<u>Circle one:</u> a) Strongly agree b) Agree c) Undecided d) Disagree 3) Strongly disagree

◆ The theory of evolution correctly explains the development of life on Earth.	a	b	c	d	e
◆ There is a good deal of scientific evidence against evolution and in favor of the Bible's account of Creation.	a	b	c	d	e
◆ There is intelligent life somewhere out there in the Universe.	a	b	c	d	e
◆ Human beings came to be through evolution, which was controlled by God.	a	b	c	d	e
◆ Humans evolved from other life forms solely through the laws of nature.	a	b	c	d	e
◆ Reincarnation really happens.	a	b	c	d	e
◆ One can believe in the Bible and Creation, *or* in atheistic evolution; there is really no middle ground.	a	b	c	d	e
◆ The first of every species (including humans) was created directly by God.	a	b	c	d	e
◆ Life began as a result of natural chemical reactions, and *not* through the power of a supernatural being or Divine intelligence.	a	b	c	d	e
◆ The Bible's account of creation should be taught in public schools as an explanation of origins.	a	b	c	d	e
◆ Life came to Earth from other worlds in the Universe.	a	b	c	d	e
◆ God created "life" in its earliest form, then His natural laws took over to eventually produce all species, including human beings.	a	b	c	d	e
◆ One can believe in scientific evolution *and* the existence of God or a Supreme Being—one belief does not exclude the other.	a	b	c	d	e
◆ Science can explain human origins, but not the origin of the human soul or spirit.	a	b	c	d	e

NOTES

1. Gallup Poll, 1990, and ABC "Prime Time," May 19, 1995; *Smithsonian Magazine*, November, 1996, p. 32.

REFERENCES

Gelles, Richard J. 1995. "Domestic Violence Factoids," University of Rhode Island Family Violence Research Program.

Gilovich, Thomas. 1991. *How We Know What Isn't So: The Fallibility of Human Reason in Everyday Life*. New York: The Free Press.

Mackay, Charles. 1980 (originally published 1841). *Memoirs of Extraordinary Popular Delusions and the Madness of Crowds*. New York: Harmony Books.

Matthews, Robert. 1998. "Blind Prejudice: Hard scientists believe they are immune to bias," New Scientist 157:12.

Pirsig, Robert. 1974. *Zen and the Art of Motorcycle Maintenance*. New York: William Morrow.

Taubes, Gary. 1998. "The (Political) Science of Salt." Science 281:898–907.

WEBSITE

Committee for the Scientific Investigation of Claims of the Paranormal (http://www.csicop.org) Site sponsored by an international organization of skeptics whose roster reads like a Who's Who of science and psychology. Covers every topic from Martians to mental telepathy; publishes *Skeptical Inquirer* magazine.

SUGGESTED READING

***How We Know What Isn't So: The Fallibility of Human Reason in Everyday Life*, Thomas Gilovich, The Free Press, New York, 1991.** Humorous examples and well-researched explanations of why we think the way we do by an expert on the psychology of belief.

Chapter 3

The E-Word

Humanity has in the course of time had to endure from the hands of science two great outrages upon its naive self love. The first was when it recognized that our earth was not the center of the universe, but only a tiny speck in a world system of a magnitude hardly conceivable. The second was when biological research robbed man of his peculiar privilege of having been specially created, and relegated him to a descent from the animal world.

Sigmund Freud

Thomas Huxley called it "the oldest of all philosophies." Today the emotional messages of *evolution* continue to incite controversy.

Larry E. is a bright thirty-something who runs a small Internet company designed to appeal to biking enthusiasts. Larry believes in using the latest technology for biking and business. But his reaction to the modern theory of human evolution may surprise you.

I don't believe animals became humans. I believe we were created from day one in the form we are now. I know scientists say monkeys changed into humans by a gradual process, but I don't see those in-between stages. I don't see half-monkeys and half-humans. And if monkeys changed to humans, why aren't there all humans and no monkeys? The fossil evidence is not convincing enough. It only raises questions. I don't think it's something that's provable. The way I believe is a faith. But scientists want to go with facts. I don't believe there are enough facts.

Larry is a creationist when it comes to human origins. He admits that what he believes is based on faith, not fact. Larry may think he remains open to scientific evidence, but his commitment to the special creation of humans will always stymie his full acceptance of evolution.

Then there's the zoo-keeper who noticed that the orangutan was reading two books, the Bible and Darwin's *Origin of Species.* In astonishment he asked the ape, "Why are you reading both those books?" The orangutan replied, "I just wanted to know if I was my brother's keeper, or my keeper's brother."

Ever since Aristotle foreshadowed Darwin's theory of evolution, the battle lines have been drawn between believers in divine creation and defenders of the natural development of all living things. To get to the raw underbelly of the evolution controversy, let's explore the fascinating but often hysterical history of the E-word.

The Greeks had a Word For It

We are list makers. It's our nature to be curious and record what we know.

Ever since human beings painted images of wild horses on cave walls, we have observed the natural world, tallied its inhabitants, and speculated on how they came to be.

Perhaps the oldest and simplest explanation for the origin of life is *spontaneous generation*—the idea that living things can suddenly emerge from non-living matter.

But even the most ordinary explanations for human origins have always been far more complex. The oldest known are variations of *creationism*, a belief still very much in evidence today. Creationism assumes many labels, but the underlying philosophy remains. Creationists demand an ultimate cause for human origins—a spiritual or supernatural force beyond the limits of our knowledge of the material world—to give purpose and meaning to life. For example:

- As early as 3500 B.C. Sumerian and Babylonian myths told of gods that had created Heaven, Earth, humans, and all creatures merely by naming them. Designation became a magical, creative act. The Hebrew book of *Genesis* followed this same format: "And God said 'Let there be light,' and there was light." (By the way, the earliest written examples of the deluge legend, which later became known as Noah's flood, are found in these same ancient texts.)
- Another common theme of human creation can be found among the Natchez, a Native American people. They believed that the first humans arose from mud animated by the breath of God.
- The idea that humans are somehow related to other animals appears in the creation myths of many primitive cultures. Tribal lore of the primitive Arunta of Australia recounts how the Self-existent Spirits swooped down from above to catch embryonic forms of life swimming in the primordial salt water shallows, then used knives to carve them into human beings. Indigenous peoples of Africa speak of their lizard pedigree. Sumatrans boast of their tiger ancestry, and Tibetans claim descent from the chattering monkey.

Today many cultures fuse spiritual images with elements from their local environment to explain the origin of humans. The Dayak of Indonesia believe that the male hornbill, a large black and white bird with a striking red beak, embodies the supreme male deity from the Upperworld. The hornbill mated with the female watersnake (or dragon) from the Underworld to produce the Tree of Life from which the first humans originated. Hornbill and dragon images feature prominently in Dayak art, and specific kinds of trees in the local environment are considered living embodiments of the Tree of Life. The dualities of the natural and the supernatural are resolved as one in Dayak culture (see Figure 3.1).

Fig. 3.1 Wooden shield from the Kenyah people of Indonesia portrays supernatural demons intertwined with spiral designs suggestive of plants and animals. (Copyright Smithsonian Institution. Enthnology catalog #301837, courtesy of Dr. Paul M. Taylor)

Over 2500 years ago the innovative thinking of early Greek naturalists opened up minds to the possibility of an earthbound answer to human origins. They wrote that nature is not ruled by the whim of gods, but has a harmony and pattern of its own that can be understood. For example:

- Anaximander speculated that life began spontaneously in mud. Fish came first, he wrote, then their descendants moved onto land and transformed over time into other animals, including humans. (Common descent)
- Empedocles preempted Darwin when he wrote that a greater variety of living things had existed in the past, but had become extinct because they lacked the skills to survive. (Survival of the fittest)

These ancient scholars had worked out two important axioms of modern evolutionary thought. But a century later the expanding perceptions of these early scientists took a detour when philosophers like Plato and Aristotle put the gods back into nature.

Plato believed that what you see is *not* what you get. He was anti-experiment, and taught that knowledge gained through intellectual insight would win out over direct observation every time. In other words, it's smarter to believe what you think than to trust what you see. Plato's own insight forced him to conclude that the various forms of life we see are only illusions—false perceptions of the ideal forms that had been created by the gods.

Plato's students really ate up all that mystical insight. They were probably as confused as ever, but were now confused at a higher level and about more important things. Plato's argument for "special creation'—the divine and separate creation of each form of life—has reverberated down through the millennia and still maintains a strangle hold on the hearts and minds of many believers.

Aristotle, a champion list maker in his own right, promoted a different view of life by listing over 500 kinds of animals and organizing them into two groups—those with and without red blood. In his *History of Animals* he describes everything from snakes and sea serpents to gall bladders and elephant tongues. He wrote that "nature proceeds little by little from things lifeless to animal life in such a way that it is impossible to determine the exact line of demarcation." Despite these inspired reflections on nature, Aristotle believed that:

1. **Life was created.** Life is the product of an intentional, supernatural act—a special creation.
2. **Life was fixed in form when created.** Life doesn't vary from those ideal forms.
3. **Each form of life is biologically disconnected from all others.** This came to be known as the "discontinuity" of the development of all living things.

These three elements—special creation, fixed forms, and discontinuity of all life—survived the Middle Ages and Darwin, and are alive and well today in many personal beliefs about human origins.

Sherry C. is a 24-year-old physical therapist whose successful practice relies on a holistic approach. Her work mirrors her belief in the unity of life, and a personal adherence to Eastern religious philosophy.

I believe that humans evolve mostly in a spiritual way. How we actually evolved physically, I'm not sure. I do believe that there was a Grand Plan, and that each kind of plant and animal was created separately. We are all created separately, then we changed. But we are all connected in a spiritual way. Our spirits are still evolving.

For Sherry, physical evolution was only a small part of where we came from. Her focus on the evolving human spirit may seem strange to those of us brought up in Western culture. But her emphasis on special creation is an ancient and common thread that still runs through a wide variety of personal beliefs about human origins.

In the 1st century B.C. the Roman poet Lucretius got things back on the evolutionary track, writing *On the Nature of Things* and speculating that a host of natural processes were solely responsible for life's diversity:

> *In the beginning the earth gave forth all kinds of herbage . . . then first put forth grass and bushes, and next gave birth to the races of*

> *mortal creatures springing up many in number in many ways . . . Then you must know did the earth first give forth races of mortal men.* (Natural origin of life)
>
> *And many races of living things must then have died out and been unable to beget and continue their breed.* (Extinction of older forms of life)
>
> *For in the case of all things which you see breathing the breath of life, either craft or courage or else speed has from the beginning of its existence protected and preserved each particular race.* (Survival of the fittest)

But with the ensuing rise of Christianity the pendulum swung back toward the scriptures. Religious doctrine dictated that every form of life had been divinely created and had not changed.

> *God created great whales, and every living creature that moveth, which the waters brought forth abundantly, after their kind, and every winged fowl after his kind . . . And God made the beast of the earth after his kind, and cattle after their kind, and every thing that creepeth upon the earth after his kind.*
>
> Genesis 1:21, 25

The word "kind" was interpreted as every unique form of life that has existed on Earth. It wasn't long before naturalists assumed that biblical "kinds" and Plato's "ideal forms" were the same thing. If you consider the old adage that animals breed true, then how do we explain the enormous diversity of life on Earth? This question remains central to the headache and heartache of the evolution controversy.

The Evolution of Evolution

In the 17th century superstition was a way of life. A common belief in vampires had originated in Eastern Europe and quickly expanded westward. Disease had been linked to corpses, but the mechanism of its spread remained unknown. Folk myth inspired science in response to the need to understand and deal with the mysteries of life and death.

Back then there were no textbooks of natural history, no sciences of biology or geology as we know them today. If you wanted to study the history of the Earth and the life on it, you had one primary source—the Holy Bible. The 17th century believer took it literally.

In 1650 Archbishop James Ussher, head of the Anglican Church in Ireland, applied his mathematical skills to a nagging question: What is the age of the Earth?

Ussher picked up his Bible and turned to the account of Creation in Genesis, then to the lists of human generations (first-born sons begetting first-born sons), and counted back from the present. From this evidence Ussher determined that the Earth had been created in 4004 B.C.—an amazing calculation that bowled over his contemporaries, most of whom were mathematically illiterate.

Then in 1654 a Hebrew scholar named John Lightfoot proclaimed that he had found further scriptural evidence that the exact day and time of the Creation had been October 23, 4004 B.C. at nine in the morning (presumably Greenwich Mean Time). Lightfoot's estimate resonated with the force of truth that only numbers can provide. Who could make up a date like that?

As the curtain rose on the 18th century, special creation and fixed forms permeated natural science. Ussher's date for the Creation enjoyed wide acceptance. It supported a young Earth less than six thousand years old, and lent an air of legitimacy to the biblical story of creation.

In all fairness, the notion that diverse forms of life are not biologically related didn't rest solely on religious conviction. Naturalists had always observed that artificial breeding had its limitations. Animals never strayed far from type. New varieties of fancy pigeons were bred every day, but no one had bred a pigeon into a chicken.

Furthermore, rock strata laid down sequentially, from the oldest on the bottom to the most recent on top, often contained sequences of fossils interrupted by large gaps. Many scientists believed the gaps would be filled as more fossils were found. Others saw them as evidence for the separate creation of each individual form of life.

But as the fervor of the Enlightenment pushed human minds beyond myth, a sea change began to take place in natural science. Pursuit of new knowledge through experiment and observation became a virtue. The word "scientist" was first put into common usage in England in the 1700s, and Darwin's theory had its roots in many innovative concepts from that era.

Geologists had long believed that the Earth was older than Ussher's claim. Layer upon layer of well differentiated rock, and evidence for dramatic upheavals that had transformed the planet's surface, argued for a history much longer than six thousand years. But most disturbing for the creationist view was the discovery of abundant evidence for extinction.

To much fanfare, the Field Museum in Chicago recently mounted an exhibit of "Sue," the world's most complete fossil skeleton of the dinosaur *Tyrannosaurus rex*. This blast from our past can trace its popularity to the Greeks and Romans, who delighted in finding fossil bones and displaying them in curio cabinets. Leonardo da Vinci was an avid collector and believed that fossils were the remains of extinct forms of animals.

In the 18th century many scientists continued to argue against the extinction of ancient forms of life because it contradicted the perfection of divine creation. They believed that living specimens of so-called extinct animals would eventually be found in unexplored regions of the world.

But in the shadow of Sacre Coeur, beneath the sidewalks of Paris, the old gypsum mines of Monmartre finally surrendered their secrets. Workmen found a profusion of fossilized bones and took them to French anatomist Georges Cuvier to be identified.

Cuvier regarded every animal as an example of the perfect integration of form and function. Because each bone was vital to the functioning of the whole organism, even small fossil fragments carried information about the entire animal. By comparing fossil bones to those of living animals, Cuvier realized that the ancient creatures found in the mines had no living counterparts. Using ingenious methods of skeletal analysis, he constructed a time sequence that demonstrated the extinction of entire groups of animals that had lived in the Paris Basin millions of years earlier.

Fig. 3.2 The Irish elk, a giant deer whose enormous antlers reached twelve feet in width, spread as far as China and northern Africa before becoming extinct 11,000 years ago. (Copyright The Natural History Museum, London)

Cuvier's groundbreaking approach to comparative anatomy earned him fame as the "father of modern vertebrate paleontology." But his most famous coup was proving the extinction of the Irish elk—not really an elk at all, nor exclusively Irish, but a giant deer whose skeleton and enormous antlers had been prized for centuries by collectors (Figure 3.2). Cuvier demonstrated that the familiar "elk" was unlike any modern deer, and could be placed in a geologic context alongside other extinct animals like the mammoth.

Cuvier had proven that an enormous diversity of life had come and gone. But the question of the source of this diversity was temporarily ignored as scientists turned their attention to the cause of extinction.

Explanations for extinction typically mirrored biblical interpretation. Fundamentalists insisted that all living forms of life were remnants of Noah's flood, the last and greatest catastrophe. But geologists quickly realized that the ready answer of a singular flood of biblical proportions wouldn't suffice. Oddities like sea shells on mountain tops, and enormous rocks transported from exotic locales, pointed to numerous ancient upheavals. Soon the "catastrophists" proposed that multiple cataclysmic events such as global floods had been responsible for extinctions. The idea grabbed hold of the scientific imagination and was soon embraced by strange bedfellows.

For example, Cuvier believed in both special creation and an Earth of immense age. He reconciled extinction and life's diversity with creation by claiming that God would supply new kinds of life after each catastrophe. The Swiss-born American zoologist Louis Agassiz, founder of Harvard's Museum of Comparative Zoology, also became a catastrophist because the theory helped him reconcile science with scripture. These men were scientists, not religious fanatics. They based their arguments on evidence, but their logic was founded on religious conviction.

Soon a geologist named Charles Lyell was to turn this illogic of biblical literalism on its head and profoundly change Earth science. Lyell argued that the steady action of observable forces such as wind and water erosion had been responsible for geologic change throughout antiquity. Slow and steady, not tumultuous and catastrophic, was Lyell's mantra. Scientists quickly rallied around this new concept and praised Lyell's victory over the catastrophists—whose scenarios had always seemed unscientific and extreme.

A little known aspect of Lyell's theory was his insistence that the Earth had not changed to any great extent since it had first formed. This forced him to use some fuzzy logic to explain how all those fossils got laid down in rocks as old as the beginning of time. However, most scientists overlooked Lyell's misstep and focused on his rational explanation for geological change that didn't invoke the scriptures. But if extinction wasn't catastrophic, as Lyell suggested, then what was the source of life's diversity?

Great minds are often so far ahead of their time that only in retrospect do we detect their brilliance. Such was the case with Georges-Louis Leclerc de Buffon, who had the audacity one hundred years before Darwin to tackle the question of life's diversity head on. In his encyclopedia of natural history, Buffon began chipping away at two millennia of dogma endorsing special creation. He discussed the similarities between apes and humans, hinted at the possibility of their common ancestry, and said organic change was induced by environmental stress. He ventured into deeper waters by suggesting that the Earth was much older than six thousand years and had been shaped and transformed solely by natural processes.

By the early 1800s, most scientists and naturalists had seen and heard enough to start questioning fixed forms. They began to study the relationships among organisms, and the role inheritance plays in passing traits from one generation to the next.

It was against this backdrop that Jean Baptiste de Lamarck, a French physician and naturalist, published the first complete theory of evolution exactly fifty years before Charles Darwin's *Origin of Species*.

Lamarck's finest contribution to biology was his division of animals into two major groups, those with and those without backbones. He coined the word "invertebrates" and single-handedly invented the field of invertebrate zoology. But Lamarck is best known for his theory of a natural mechanism for organic change—a wonderful example of scientific thinking that seems ridiculous to us today.

In Lamarck's view, all life began as simple, undifferentiated blobs of life that were the result of special creation. (It just doesn't go away.) These blobs progressed in complexity and perfection through a mechanism called "inherited adaptation"—commonly lampooned using the example of a giraffe reaching higher and higher into the tops of trees for leafy shoots. As the animal repeatedly stretches to eat, its neck grows longer like Pinocchio's nose. This longer neck is passed on to its offspring. The giraffe's interaction with its environment results in a new and beneficial trait which can be inherited.

This part of Lamarck's theory made sense to many people, despite the fact that centuries of cropping the ears and tails of dogs had done nothing to change the ears and tails of their offspring. However, Lamarck's unabridged theory of evolution was harder to swallow. He said organisms aren't passively altered by their environment. They actively modify their behavior, favoring the use or disuse of certain body parts. This leads to an increase or reduction in size, and even a loss of some features. In other words, "use it or lose it" and the result will be inherited. Lamarck believed that this mechanism of slow, gradual change, driven by the animal's interaction with its environment, generates new forms of life.

Lamarck's theory sounds remarkably similar to modern evolution. But the two differ in crucial details. Lamarck emphasized a directional force in evolution from simple to more complex and more perfect. He believed that organisms have an inner will or sense of purpose that compels them to change. Lamarck also rejected extinction, claiming that varieties of life disappear because each completely transforms into a different one. Lamarck's evolution was one dimensional. Lines of descent were isolated and didn't branch.

By the mid-1800s over two thousand years' worth of natural science was finally poised to collide with creation and religious fundamentalism. Into that rich stew of fanciful theories and puzzling evidence rode a man who threatened to resolve the question of biological diversity once and for all. Charles Darwin would dazzle the world with his multidimensional scheme of evolution that produces both the warp and woof in the fabric of life. As new kinds of animals branch off from existing ones, common descent weaves the hereditary connection among all life. Evolution would become purposeless, directionless, and unlimited.

What is a Species, and Why the Fuss?

It all comes down to this. When you pare it down to its bare essentials, silence the rhetoric, and move to the bottom line, this is where you'll be. It's the spot where the rubber meets the road in the evolutionary controversy. It starts with "S"—and that stands for "Species."

Today the most widely accepted definition of a species goes like this:

> *A group of plants or animals whose members share certain traits and breed exclusively among themselves.*

Since ancient times naturalists had grouped together plants and animals that shared the same characteristics. They recognized a limited amount of variation within each group. But they didn't think these small differences were responsible for the larger, more defining differences that separated the biblical "kinds"—for example, reptiles and birds, or dogs and cats. They had also noticed traits common to all living things, but didn't see them as evidence for a biological connection. Theories like Lamarck's were often ridiculed because naturalists saw no need to explain the appearance of new species. Each had been a special product of God's plan. Extinctions were just speed bumps along the divine highway.

Even the most primitive cultures and savage minds are compelled to identify and name the living things in their environment. Scientists who study this very basic human activity call it *folk biology*. It's a survival instinct that turns fear of the unknown into knowledge and power. Cultures transmit this knowledge and power to future generations. There is persuasive evidence that the sorting and naming of plants and animals found in the folk biology of indigenous cultures correlates to a high degree with our Western classification of species.

Smithsonian anthropologist Paul Michael Taylor conducted one of the most comprehensive studies of the folk biology of a single human culture, that of the Tobelo people on Halamera Island in Indonesia. Tobelo subsistence technology, material culture, and medicine rely upon a complex classification of plants and animals in their environment. They have developed their system over untold generations within the unique cultural context of Tobelo language, local ecology, and value of each plant and animal to their survival. Many plants and animals that are given special names by the Tobelo have been identified by Western biologists as separate species.

Ernst Mayr, considered by many to be the greatest evolutionist of the 20th century, relates his own experience with folk biology while living among the tribes of Papua, New Guinea, writing that "these superb woodsmen had 136 names for the 137 species of birds I distinguished. That Stone Age man recognizes the same entities of nature as Western university-trained scientists refutes rather decisively the claim that species are nothing but a product of the human imagination."

Scientists estimate that there are over ten million species of animals. Each year new species are discovered. For example, in 1995 an expedition to an isolated region of northeastern Tibet discovered the Riwoche horse, a new species that traces its lineage back five million years. The archaic looking horse is short, beige in color and resembles the horses in prehistoric cave paintings. Scientists believe early cave dwellers hunted it and ate it, but only in the last three to four thousand years has it been domesticated.

Most scientists accept that a species is a concrete unit in nature. As a definable product of evolution, species is a small but powerful concept. Every existing species has gone through evolution's filter and is a masterpiece of survival.

A species tends to be stable and self-perpetuating. Because its members only breed among themselves, beneficial traits aren't diluted out by genes from other species, or mixed with harmful mutations. But it's the emergence of new species that drives evolution.

The modern theory of evolution contends that genetic variations among individual members of a species will accumulate if they help those individuals survive and reproduce. Small genetic differences can make a big impact on survival rates. For example, individuals with a slightly different form of digestive enzyme could eat and process scarce nutrients more efficiently in times of famine. As their survival rate increases they are more likely to pass on their unique differences to offspring. Their new ability distinguishes them genetically, and perhaps physically, from the rest of the group. If they become reproductively isolated because of geography or other barriers to mating, they may split off from the larger group and become a new species.

Today most of the controversy over evolution revolves around the origin of animals that differ by several orders of magnitude beyond a species. For example, evolutionists tell us hippos and whales share an ancient common ancestor, a land mammal whose descendants moved back into the water and evolved flippers from legs. This may sound like Ripley's "Believe It or Not," but there is convincing fossil and genetic evidence that this occurred. If you are a creationist, you might dismiss the evidence out of hand. Or you might be one of those people who believe in the power of evolution up to a point, but just can't quite accept such a far reaching premise. You may not be a creationist, but just skeptical and want to see evolution's smoking guns.

On the other hand, you may believe that all life evolved naturally with the exception of human life. Evolutionists tell us we are all one species, *Homo sapiens*. We also belong to the same family that includes the living great apes—gorillas, orangutans, and chimpanzees. We know that our genes differ by less than 2% from those of the chimpanzee. This very small divergence in our genes has resulted in crucial differences. How is that possible?

Darwin didn't use the E-word to describe his theory because common usage in the 1800s gave *evolution* a very different meaning from the one it has today. Its Latin root (*evolutus*) means "rolling out from the center of a circle," like the unrolling of a piece of carpet. Its use would have implied that life is on a prescribed, preordained path—a "Divine Rollout" that Darwin didn't want associated with his theory. In modern jargon, *evolution* has come to mean just the opposite—unlimited, random hereditary change constrained only by natural forces.

Smashing the Pedestal of Perfection

If you accept that species evolve in the modern sense, then all life is one.

We are all made of the same stuff, linked by change and common descent, and connected by a biological history of billions of years. The human species got here the same way all the others did. There is no progress of life from less to more perfect that ends with human beings at the top.

How do you like being in that position? Does it make you feel any less unique, intelligent, or important? For many people, it does. In the final analysis, the crux of the evolution-creation argument is not about apes, but ego.

Freud's tongue-in-cheek lament that "biological research robbed man of his peculiar privilege of having been specially created, and relegated him to a descent from the animal world" belies the fact that Freud applied evolutionary concepts to psychoanalysis whenever possible. He empathized with Darwin's experience and compared it to his own:

> *Man's craving for grandiosity is now suffering the third and most bitter blow from present day psychological research which is endeavoring to prove to the 'ego' of each one of us that he is not even master in his own house.*
>
> *We psycho-analysts were neither the first nor the only ones to propose to mankind that they should look inward; but it appears to*

> *be our lot to advocate it most insistently and to support it by empirical evidence which touches every man closely. This is the kernel of the universal revolt against our science.*

Darwin's theory made a powerful but threatening argument for a new way to view the world and our place in it. It's not divine magic but magnificent Nature that has placed all creatures on the level playing field of life. The human species has earned its status on that field through millions of years of struggle and survival. We could also lose it very quickly.

Sixty-six years after Darwin published his theory of evolution, a backlash of Victorian traditionalism against the modernism and intellectual experimentation of the "roaring twenties" spawned an American cultural phenomenon known as Christian Fundamentalism. Though mainly based in the conservative religious sects of the South, this movement became international news when a small town in Tennessee became host to the courtroom battle that literally put Darwin on trial.

Fiery Pulpits: Tennessee v. John Thomas Scopes

George Rappalyea is not exactly a household name. And Dayton, Tennessee seems an unlikely place for a conspiracy. But without Rappalyea and Dayton there would never have been a "Monkey Trial."

In the summer of 1925, Dayton was a quiet, prosperous community nestled in the foothills of the Smokey Mountains, mid-way between Knoxville and Chattanooga. Many of its eighteen hundred citizens farmed tobacco, wheat, or soybeans. Most were Bible-toting Christians.

George Rappalyea was a 31 year old geologist from New York who had come to Tennessee to survey and had discovered a large coal deposit near Dayton. He stayed on as manager of the Cumberland Coal and Iron Company.

Rappalyea soon developed a strong dislike for the hellfire and brimstone moralizing of the local Christian Fundamentalists. When Tennessee passed an Anti-Evolution Law in early 1925 that prohibited the teaching of evolution, Rappalyea decided to do something about the Fundamentalists and their law.

Rappelyea hatched his plot in Robinson's Drugstore, the local gathering spot for Dayton's movers and shakers. There Rappelyea convinced town leaders that a suit against the evolution law would put Dayton on the map and bring new residents to the declining town. He should have heeded the warning, "Be careful what you wish for because you just might get it."

Rappalyea knew that the American Civil Liberties Union had offered to defend any teacher charged with violating Tennessee's Anti-Evolution statute. When his friend John Scopes, the local high school science teacher, joined the discussion at Robinson's and commented that it would be impossible to teach science without teaching evolution, Rappelyea knew he had found his mark.

Scopes admitted to the crowd at Robinson's that he had violated the law by assigning readings on evolution to a biology class using a state approved text, *Hunter's Civic Biology*. When asked by Rappalyea whether he'd be willing to take part in a test case, Scopes agreed. Rappalyea swore out a warrant and the Dayton police arrested and indicted Scopes four days later (Figure 3.3).

Fig. 3.3 John Scopes was a twenty-four-year-old college graduate, science teacher, and willing defendant in America's first court case pitting evolution against creationism.
(Courtesy Bryan College Archives)

The red-haired, boyish looking Scopes became a willing defendant in the case. He never testified during the trial, having readily admitted that he had broken the law by teaching Darwin's theory of evolution. But several of his students, including fourteen-year-old Howard Morgan, were called to testify as witnesses for the prosecution. Young Morgan related that Scopes had taught him that after the Earth had cooled, "there was a little germ of a one-celled organism formed, and this organism kept evolving until it got to be a pretty good-sized animal and then came to be a land animal, and it kept on evolving and from this was man, and that man is just another animal."

The ACLU's legal team starred Clarence Darrow, America's most famous defense attorney. Darrow was still basking in notoriety from his 1924 hangman's reprieve for Nathan Leopold and Richard Loeb, two teenagers convicted of the kidnap and murder of Bobby Franks.

For the first time in his career Darrow volunteered his services and agreed to defend Scopes. The ACLU had reservations about the liberal, agnostic Darrow, fearing he would turn the trial into an all-out assault on religion instead of an attack on the constitutionality of Tennessee's anti-evolution law. But Scopes wanted Darrow, and the rest is history.

The prosecution's team also boasted a celebrity volunteer. William Jennings Bryan, famed orator and three-time loser on the Democratic Presidential ticket, had not practiced law for nearly thirty years. Instead the "Great Commoner" had used his popularity and persuasive rhetoric to support the fundamentalist crusade against evolution. His repeated comments on evolution ("When I want to read fiction, I don't turn to the Arabian Nights; I turn to works of biology. I like my fiction wild.") and his evaluation of its worth ("The Rock of Ages is more important than the age of rocks.") soon earned him the nickname "the Fundamentalist Pope."

Famed writer and social critic H.L. Mencken, whose employer the *Baltimore Sun* helped bankroll Scopes' defense, was one of many reporters who came from as far away as London to cover the trial. In describing Bryan's simple nature Mencken said, "He was probably the least sophisticated of American public men. He saw life as uncomplicated, for he ignored its complications. Anything that was not clear to him he treated as untrue."

The townspeople of Dayton warmly embraced Bryan. They fed him and bought his book *Hell and the High School*, a scathing case against the teaching of evolution in public schools. As the trial got underway in the sticky July heat, Bryan's book was sold alongside lemonade and hot dogs. Chimps performed and banners flew in the town's carnival atmosphere (Figure 3.4).

Fig. 3.4 The sidewalks of Dayton, Tennessee during *Tennessee v. Scopes.* (Courtesy Bryan College Archives)

Scopes was tried in an atmosphere reminiscent of a Billy Sunday revival. On July 10, 1925 Judge Raulston opened the trial with a prayer.

The prosecution offered six days of testimony that verged on sermonizing, accompanied by frustrated outbursts by the defense team and stubborn posturing on both sides. The proceedings did little but establish Scopes' teaching of evolution as deserving of hell fire and damnation should it continue.

As the defense rose to present its case, Judge Raulston refused to allow testimony on evolution from Darrow's panel of eight expert witnesses. They included the head of the geology department at Harvard, several renowned biologists, zoologists, and educators, and the geologist for the State of Tennessee. But the scientists' depositions were read into the trial record anyway and printed by newspapers around the world. Darrow had made his point.

John Washington Butler, the tobacco farmer and Tennessee legislator who wrote the anti-evolution law exactly as passed, reportedly favored hearing testimony from the experts, saying,

> *The judge ought to give 'em a chance to tell what evolution is. Course we got 'em licked anyhow, but I believe in being fair and square and American. Besides, I'd like to know what evolution is myself.*

On the seventh day before they rested, the defense called opposing counsel William Jennings Bryan to the stand. Few judges today would permit this type of grandstanding, but without witnesses to present and deprived of the usual defense strategy, Darrow had no choice.

By now the courtroom crowd had swelled into the thousands. Judge Raulston moved the trial outside to the courthouse lawn, fearing that the courtroom floor would collapse under the weight of the mob. A sign on the courthouse wall commanding "READ YOUR BIBLE" in foot-high letters hung ominously close to the jury, but Darrow was successful in having it removed before questioning began (Figure 3.5).

Bryan should have let Darrow have his expert witnesses. Instead, he readily accepted Darrow's challenge as a chance to defend both the Bible and his own faith. What followed was a landmark in the annals of American jurisprudence. Here is an excerpt from Darrow's interrogation of Bryan to give you a sense of the moment. Darrow questions Bryan on the logic and plausibility of biblical accounts:

Fig. 3.5 John Scopes (left) with his initial defenders, lawyer John Neal (middle) and George Rappelyea (right), beneath a trial-related poster. (Courtesy Bryan College Archives)

Darrow: The Bible said Joshua commanded the Sun to stand still for the purpose of lengthening the day, doesn't it, and you believe it?

Bryan: I do.

Darrow: Do you believe at that time the Sun went around the Earth?

Bryan: No, I believe that the Earth goes around the Sun.

Darrow: Do you believe that the men who wrote it thought that the day could be lengthened or that the Sun could be stopped?

Bryan: I don't know what they thought.

Darrow: Don't you believe that in order to lengthen the day it would have been construed that the Earth stood still?

Bryan: I would not attempt to say what would have been necessary, but I know this, that I can take a glass of water that would fall to the ground without the strength of my hand, and to the extent of the glass of water I can overcome the law of gravitation and lift it up; whereas, without my hand, it would fall to the ground. If my puny hand can overcome the law of gravitation, the most universally understood, to that extent, I would not set power to the hand of Almighty God that made the Universe.

Darrow: Now, Mr. Bryan, have you ever pondered what would naturally happen to the Earth if it stood still suddenly?

Bryan: No.

Darrow: Don't you know it would have been converted into a molten mass of matter?

Bryan: I would want to hear expert testimony on that.

Darrow continued to pummel Bryan with questions on Noah and the great flood, Eve's temptation of Adam in the garden of Eden, and the story of creation in Genesis, as he chipped away at Bryan's logic and ability to defend a literal reading of the Bible (Figure 3.6).

Fig. 3.6 Judge's view of Clarence Darrow (right) interrogating William Jennings Bryan (left) on the courthouse lawn in the July heat of 1925. (Courtesy Bryan College Archives)

At one point the increasingly irritated and flustered Bryan retorted, "I do not think about things I don't think about." Darrow asked, "Do you think about the things you do think about?"—and Bryan replied, "Well, sometimes."

Testimony ended with Bryan accusing Darrow of attempting to "slur at the Bible." Darrow countered, "I object to your statement. I am examining you on your fool ideas that no intelligent Christian on Earth believes."

The next day, Judge Raulston struck Bryan's testimony from the record, but it had already been widely reported as a humiliation and defeat. H. L. Mencken reported, "Darrow has lost this case, but it seems to me he has nevertheless performed a great public service by fighting it to a finish and in a perfectly serious way. It serves notice on the country that Neandertal man is organizing in these forlorn backwaters of the land, led by a fanatic, rid of sense and devoid of conscience."

The trial lasted eleven days, and at the end Darrow asked for a directed verdict of guilty so he could start the appeal. The jury agreed and the judge fined Scopes one hundred dollars. Two years later the conviction was overturned on appeal and dismissed by the State Supreme Court. Scopes never resumed teaching, but went back to school and became a geologist, like his friend Rappalyea. Bryan, who was seventy and in poor health during the trial, died shortly thereafter.

Without George Rappelyea there would not have been what many have termed the trial of the century. Like a ghostly litigant, the Monkey Trial still haunts American courts as a rallying point for the creationist cause.

Tennessee repealed its anti-evolution law in 1967, and in 1968 the Supreme Court of the United States held unconstitutional any statutes prohibiting the teaching of evolution in public schools. In 1987 the Supreme Court decision in *Edwards v. Aguillard* found unconstitutional a Louisiana law *requiring* creationism to be taught along with evolution in public schools.

But efforts continue to legitimize the teaching of religion-based alternatives to evolution. As recently as 1999 the Kansas

State Board of Education voted to delete questions about evolution from its standardized high school tests.

If you attended high school in the 1960s, you may not have been exposed to Darwin's theory, or your teachers may have covered it only in a tentative way. Despite legislative efforts to ensure the teaching of evolution as science, it carried with it a taint of the immoral and thus inched its way slowly into the curricula of even the best public schools.

Fully 60% of my Belief Survey participants report that they were taught evolution in high school science classes—not surprising since over 65% of them were twenty-nine years of age or younger. About the same percentage support the teaching of evolution, not biblical creation, as an explanation for origins.

The fundamentalist-modern clash just won't go away. It's all about putting Humpty-Dumpty back up on the pedestal of perfection. Now let's take a look at the man who knocked him off.

REFERENCES

Campbell, Joseph. 1973. *Myths to Live By*. New York: Bantam.

Freud, Sigmund. 1920c; reprint 1935. *A General Introduction to Psycho-Analysis*. New York: Liveright.

Gould, Stephen Jay. 1980. *The Panda's Thumb*. New York: W.W. Norton.

Mayr, Ernst. 1963. *Animal Species and Evolution*. Cambridge: Harvard University Press.

Taylor, Paul Michael. 1990. "The Folk Biology of the Tobelo People," *Smithsonian Contributions to Anthropology* 43. Washington, DC: Smithsonian Institution Press.

Taylor, Paul Michael and Lorraine V. Aragon. 1991. *Beyond The Java Sea: Art of Indonesia's Outer Islands*. Washington, D.C.: Smithsonian Institution and Harry N. Abrams.

WEBSITE

Famous Trials in American History (www.law.umkc.edu/faculty/projects) Award-winning site created by Professor Douglas Linder at the University of Missouri - Kansas City Law School. (At Website above, click on Linder's Famous Trials Page) Uniquely entertaining, well-documented goldmine of information on the Scopes trial and other landmark cases. Includes bios, trial excerpts, and social and political commentary.

SUGGESTED READING

Ernst Mayr, *One Long Argument: Charles Darwin and the Genesis of Modern Evolutionary Thought*, Harvard University Press, Cambridge, 1991. Readable, concise account of the development of Darwin's theory. Mayr clarifies the buzz words and puts the evolution-creationism debate in scientific context.

Chapter 4

The Great Darwin Divide

A curious aspect of the theory of evolution is that everybody thinks he understands it.

Jacques Monod, Nobel prize-winning biochemist
On the Molecular Theory of Evolution, 1974

Charles Darwin was a pack rat.

When he died in 1882, he left bird nests, pebbles, thousands of prized beetles and barnacles, and a giant sloth tooth from South America. He had saved every letter, diary, and notebook from his seventy-three years. In addition over four thousand letters written by Darwin had been saved by admiring friends and relatives.

Through this detritus of a brilliant career we glimpse the unabridged Darwin—a complex man leading a contradictory life. Though raised in church-influenced gentility, Darwin becomes a reluctant dissenter. We sense his apprehension, then elation, as he grasps the significance of his ideas. The man who couldn't throw out a scrap of paper would labor his entire life to rid his world of its ruling orthodoxy.

Darwin's argument against divine creation changed forever the focus of Western thought. Despite one-hundred-fifty years of refinement and intense debate, his theory has stood the test of time and remains the most widely accepted scientific explanation for human origins. It is also the most controversial doctrine in the history of science.

Would you answer "True" or "False" to each of the following statements?

- Charles Darwin originated the phrase "survival of the fittest."

- Darwin said that the role of natural selection is to eliminate the weak.
- Evolution means progress toward perfection.

If you answered True to any of the above, then you only think you know what Darwin said. Words like *evolution* and *natural selection* carry a high recognition factor, proof of Darwin's enduring legacy. But British philosopher Herbert Spencer was the first to popularize the phrase "survival of the fittest." Spencer used it to describe the competition among individuals and nations that led, in his view, to wealth, power, and the excesses of Capitalism. Today we use "survival of the fittest" to describe activities ranging from mountain climbing to stock trading. Darwin didn't say it, but the words are synonymous with his ideas.

Edward Blyth, a naturalist and creationist, first used the term "natural selection" to describe a conservative force in nature that weeds out the unhealthy and scraps traits that deviate too far from the ideal blueprint laid out by an Intelligent Designer. In contrast, Darwin said natural selection builds new species and is a creative force in nature.

Finally, Darwin said evolution doesn't have a goal of perfection. Evolutionary success simply means the ability to change and adapt to environmental stresses in order to survive and reproduce.

Jennifer S. is a well-educated single parent in her thirties who works as an editor. She readily admits she never thought a lot about evolution until she became an adult.

I believe in evolution, but I'm not clear on all the specifics. I don't buy into the Adam and Eve thing. I really never thought about it until I was older. You might say I lost faith, and rather than losing anything I felt freed. I think people grow intellectually by not being circumscribed by a religious belief.

I guess I would say that coming up through time animals changed little by little as the climate changed. And the strong killed off the weak. I think they have found that evolution doesn't necessarily fit Darwin's theory in many cases, like its not always a steady progression upwards.

But I think humans are at the top. We're the most perfect. With our brains and how we change our environment, it's hard to imagine what the next step would be. I don't think we're still evolving. I think we're it.

Jennifer's blind commitment to the concept of evolution mirrors that of many people. They support Darwin's theory despite a vague understanding of its details. Their loyalty is admirable, but common misconceptions about evolution can make it vulnerable to crackpot theorists and inroads from the non- science of "creation science."

Was Darwin's theory truly revolutionary? Why does it remain controversial? Does the evidence support Darwin's speculations? Was Darwin right?

True Tales From The South Pacific

It all began with twenty years of silence.

In the mid 1800s a young British naturalist sailed halfway around the world to collect plants and animals in South America and on the islands of the South Pacific. He was dazzled by the enormous variety of birds, reptiles, and insects he found there, and became convinced that special creation couldn't account for such an extraordinary amount of variation. Not only were the animals adapted to unique island habitats, but they also closely resembled their mainland counterparts.

Soon he began to question common wisdom. Surely species can change, and generations of change could produce new species. The ideal match between each animal and its habitat suggested that the environment could act as a catalyst for change. In 1855 he compiled his thoughts and published them in a paper titled "On the Law that has Regulated the Introduction of New Species." During the next three years he searched for the natural processes that would generate new species.

He read Thomas Malthus's "Essay on the Principle of Population" that describes the natural struggle for survival in a healthy population when food and resources become scarce. It's winner take all, and survivors are those best equipped to grab the goods and reproduce. They are the fittest. This natural competition ensures a healthy population.

He quickly realized that every island species was a survivor of this same kind of struggle, and speculated that:

- For any given species, more offspring are born than can survive.
- In the random shuffle of heredity, each individual is born slightly different from the rest.
- Differences in body and behavior can determine survival.
- Only the fittest individuals will adapt, survive, reproduce, and pass on their special abilities to their offspring.
- Over the course of many generations, unique combinations of the most beneficial traits will accumulate and form new species.

The young naturalist was Alfred Russel Wallace. In 1858, armed with his new insights, Wallace drafted a detailed theory on the origin of species that he called "natural selection" and mailed his manuscript to fellow naturalist Charles Darwin. (Figures 4.1 and 4.2)

On the surface, Darwin and Wallace could not have been more different. Yet there were subtle similarities. Born fourteen years before Wallace, Charles Darwin was raised in an upper-class English family that gave him both the means and opportunity for intellectual pursuit. In contrast, Alfred Wallace was born into a troubled family that teetered on the brink of poverty. Unlike the Cambridge-educated Darwin, he was a self-taught, self-made man who trained himself to be a surveyor. This led him into the new field of geology where he discovered the wonders of fossils. Like Darwin he became an avid naturalist and collector.

Fig. 4.1 Charles Robert Darwin, naturalist credited with developing the theory of natural evolution which provides the foundation for modern genetics and biological science. (Copyright The Natural History Museum, London)

Fig. 4.2 Alfred Russel Wallace, co-discoverer of the theory of natural selection and evolution. (Courtesy Dr. Charles H. Smith)

For many years the paths of these two men with their shared passion for nature and discovery never crossed. In 1831 the 22-year-old Darwin set off on the ten-gun brig H.M.S. *Beagle* on a voyage of personal discovery which took him to the coasts of

South America and the Pacific. For the next five years he collected plants and animals from the Amazonian rain forests and pried fossils from ancient rocks in the Andes. An inveterate researcher, he filled eighteen notebooks with his observations and sketches.

At the start of the voyage Darwin had held the same belief as most naturalists of his day that species had been divinely created and fixed in form. Like Wallace, he was overwhelmed by the enormous variety of plants and animals he found on his journey and by mid-voyage had become haunted by the nagging suspicion that species could change. He had read Lyell's history of the Earth and theorized that during those millions of years Nature's ingenuity had multiplied a handful of species into the abundance he was seeing.

Almost as an afterthought, Darwin hastily collected several new kinds of finches native to the Galapagos Islands that bore a puzzling resemblance to mainland finches. When he returned to England, a noted ornithologist confirmed that the finches were unknown species. It was already clear to Darwin that the Galapagos finches had descended from mainland species. It was only a few short mental steps from there to the possibility that all finches had descended from a common finch ancestor, and that all higher organisms had descended from a few ancestral forms. At that point Darwin had arrived at the first key element in his theory —Change and Common Descent.

By 1838 Darwin had developed the second key element—Natural Selection—which he described in his diary:

> *In October 1838, fifteen months after I had begun my systematic enquiry, I happened to read for amusement Malthus on Population, and being well prepared to appreciate the struggle for existence which everywhere goes on . . . it at once struck me that under these circumstances favourable variations would tend to be preserved, and unfavourable ones to be destroyed. The result of this would be the formation of new species.*

Then there was silence. With the exception of a short unpublished paper and a longer essay, Darwin published nothing further on his theory.

Meanwhile Wallace eagerly devoured Darwin's *Journal of the Beagle.* He became so enamored of Darwin's description of plants and animals in South America that he sailed to the Amazon in 1848. Eventually he made his way to the spice islands of Malaysia where he survived on a shoestring, content to live a simple, rugged life among the natives. While recovering from one of his many bouts of fever, Wallace began to speculate about the source of life's diversity. Why would the Creator have invented so many different versions of the same organism for basically the same environments? Like Darwin he had read Lyell's theory that slow, natural changes had shaped the Earth. Wallace now applied the same reasoning to the development of life. There had been no divine magic involved.

Insights gained during eight years in the tropics inspired Wallace to independently propose a theory of evolution virtually identical to that of Darwin. But today we talk about Darwin's theory not Wallace's. Darwin's ideas catapulted him to fame and immortality, while Wallace exited to the wings.

Into the Fire

There are as many theories about Darwin's long silence as there are biographies of the man. Chronic ill health plagued him most of his life. He sailed on the *Beagle* as an energetic, adventurous young man. But shortly after returning to England he began to complain of nausea, insomnia, and stomach pains. He became more anxious, depressed, and socially withdrawn. The slightest stress magnified his symptoms.

In 1842 at the height of the social and political turmoil of Victorian England, Darwin and his wife Emma escaped London for the quiet of their country estate in Kent. The move gave Darwin the social isolation he craved. Protected from intellectual distraction, he completed a study of barnacles, bred pigeons, and gathered data to support his emerging theory of the natural origin of species.

During those years Darwin led two lives—wealthy country landowner and closet evolutionist. His poor health continued as he tried fad diets and spa cures to no avail. Scholars have always assumed that Darwin suffered from a tropical disease he had picked up on the voyage to South America. But evidence from Darwin's diaries has recently led some experts to suspect there was a more familiar cause. When considered separately, Darwin's symptoms indicate a variety of ailments. But taken as a whole they constitute nine out of the thirteen diagnostic signs of panic disorder. Darwin was able to live an outwardly normal and productive life by avoiding crowds, carrying on a prolific correspondence, and keeping stress to a minimum. This was not a man who would welcome public controversy.

Darwin remembered the public outcry against the book *Vestiges of the Natural History of Creation*, published anonymously by Robert Chambers in 1844. Chambers was an amateur scientist who had developed his own theory of evolution—a mixed bag of spontaneous generation, inheritance of acquired characters, and vague "inner urges to evolve." Lacking a coherent mechanism for his theory, Chambers fell back on the crutch of creation. He claimed that the hierarchy of life from simple to complex reveals divine programming. After the laughter died down, clergy and scientists alike denounced the book. Darwin feared that his own ideas would brand him as a heretic and atheist.

It is more likely that Darwin's long silence resulted from a combination of many factors residing in a naturally timid man. His chronic ill health made him a recluse, restricted his energy, and thwarted his will to jump headlong into controversy.

By 1858 Darwin had managed to complete nine or ten chapters of his "big species book" when Wallace's manuscript on natural selection arrived. It wasn't the first time the two men had corresponded. Darwin had read with some interest Wallace's earlier paper. But he had viewed it merely as an inconvenience, and had brushed off Wallace like a pesky fly. After all, Wallace was a virtual unknown who had taken only five years to come to the same conclusions that Darwin had been pondering over for nearly twenty.

But this time Darwin panicked. He could see his life's work going down the drain. He sought the advice of fellow scientist Charles Lyell, and the botanist Joseph Hooker. They persuaded him to let them present his unpublished theories, together with Wallace's paper, at the Linnaean Society in London—the 19th century equivalent of a joint press release. Meanwhile Darwin stayed at home, writing feverishly.

Finally in 1859 a smaller version of Darwin's big species book was hurriedly published as the *Origin of Species*. On the day of its publication the entire first printing of 1,250 copies sold out.

Darwin's blockbuster portrait of the natural evolution of life came as a breath of fresh air for many Victorian intellectuals. He had seen no need to invoke divine power or the supernatural. Instead Darwin built his case on observation, logic, and tangible evidence.

Within ten years of the publication of the *Origin of Species* most scientists had accepted the idea of change and common descent. But natural selection, the element in Darwin's theory most useful to naturalists, had to wait another seventy-five years to gain the blessing of science.

Pieces of the evolution puzzle had been around since the early Greeks. Aristotle had written that "nature proceeds little by little from things lifeless to animal life." Lamarck had theorized that acquired change can be inherited. Even the English creationist Bishop William Paley embraced a role for adaptation because he thought it reflected the structure of divine thought. But these stale notions endorsed a static hierarchy of life and an inevitable progress toward perfection.

Darwin's genius lay in his ability to think out of the box. He seized old theories and severed them from scripture. He let his creative juices flow. Like Sherlock Holmes, he eliminated the impossible and accepted what remained as truth—no matter how improbable. Bound only by observation and evidence, he modified and molded each piece of the puzzle, lifting the fog of twenty-five hundred years of natural theology. (Figure 4.3)

Fig. 4.3 Pieces of the evolution puzzle connected by Darwin's theory.

What Darwin Said—and Didn't Say

In a nutshell, Darwin said,

I am fully convinced that species are not immutable; but that those belonging to what are called the same genera are lineal descendants of some other and generally extinct species. Furthermore, I am convinced that Natural Selection has been the most important, but not the exclusive, means of modification.

It was a deceptively simple yet powerful scheme. Let's break it down into five interlocking elements:

- **Change and Common Descent**—the *what* of evolution
- **Natural Selection**—the *how* of evolution

- **Variation**—evolution's *raw material*
- **Speciation**—the *product* of evolution
- **Gradualism**—evolution's *tempo*

Darwin's paramount goal was to establish *evolutionary* change as fact. This wasn't plain old change. Lamarck had been there, done that. Darwin called his new idea of change "descent with modification" or **Change and Common Descent**, implying the biological connection of all life on Earth.

Darwin said life is fluid, not static. Species change and *branch* into new kinds of life. New traits arise randomly among individuals and are passed to their offspring. Over time these modifications accumulate, fashioning new organisms that no longer resemble their ancestors. Most examples of evolution in the *Origin of Species* illustrate the branching biological change that was vital to Darwin's argument. This view of evolution as a creative force defined Darwin's theory and distanced it from Lamarckism, creationism, and all the other "isms" that were constrained by ideal forms and fixed species.

Ever since Darwin the tree of life has implied biological kinship. Darwin used fossil evidence to follow the family tree of species back in time, through parent species then grandparent species, until multiple lines resolved into a common ancestor. Today evolutionists follow that same trail of evolution by comparing molecular similarities among the genes and proteins of all living things.

Natural Selection was Darwin's engine of evolution. The biologist Julian Huxley wrote, "Natural selection rendered evolution scientifically intelligible."

Darwin knew a good thing when he saw it. Nearly overnight Lyell's theory had multiplied the age of the Earth by a thousandfold, and scientists were using those millions of years of natural change to explain baffling geological phenomena. Darwin believed that as global environments had changed, local habitats would also have been altered. Along with nature's remodeling came new stresses on organisms. Survival would have depended on individual skills and traits that supported adaptation.

> *If then we have under nature variability and a powerful agent always ready to act and select, why should we doubt that variations in any way useful to beings, under their excessively complex relations of life, would be preserved, accumulated, and inherited?*
>
> *What limit can be put to this power, acting during long ages and rigidly scrutinising the whole constitution, structure, and habits of each creature,—favouring the good and rejecting the bad? I can see no limit to this power.*

Darwin said natural selection is the unlimited architect of life. He challenged Blyth's view that natural selection only weeds out mistakes to maintain the divine status quo. Instead Darwin characterized natural selection as a creative feedback system that tolerates the better adapted and discards ineffective designs. No two people, not even twins, are identical in every detail. The key to nature's improvisation via natural selection is this constant flow of subtle, random differences among individuals of any species.

But Darwin also considered nature a little sloppy. He believed natural selection was the most powerful force for evolutionary change, but not the only one. A beneficial change in one area can cause an undesired effect in another. And some traits that are key to survival may also be able to carry out less essential functions, thus pulling double duty along evolution's trail. Most people equate natural selection with *survival of the fittest*. But Darwin didn't say the fittest are the best, only the best adapted.

Each winter we experience the effects of natural selection on a very personal level. A new strain of flu virus makes the rounds in offices and classrooms, infecting some of us and sparing others. Our natural ability to ward off disease relies on an inherited immune system that varies in each of us.

Darwin anticipated objections to his concept of natural selection and devoted an entire chapter to answering them. Despite these efforts the attacks against natural selection continued as scientists debated the precise manner in which change and variation occur.

Darwin struggled throughout his life with the source of **Variation**, evolution's raw material. He knew first hand the difficulties of breeding variety into domestic animals. Some results can never be achieved. Artificial matings often produce non-viable or sterile offspring. For example, mules are the sterile offspring of a horse and a female ass.

Darwin also ascribed to a popular theory called "blending." For example, mating a black dog with a white dog was thought to produce gray puppies because the source of the black or white color in the parents becomes blended in the offspring. But taking this concept a few steps further, all traits including those that favor survival would quickly disappear. Darwin's belief in blending stymied his search for the source of variation.

Both Darwin and Wallace believed that variation must be innate and present in the organism at birth. Unlike Lamarck, Darwin said nature merely acts upon traits already present, and doesn't induce new ones. How does this inborn variation occur? Darwin suggested that "a change in the conditions of life, by specially acting on the reproductive system, causes or increases variability." (Environment causes variation.) Then he hedged, saying that "in looking at many small points of difference between species, which seem to be quite unimportant, we must not forget that climate, food, etc., probably produce some slight and direct effect." (Environment *might* cause variation.)

Finally, in frustration, Darwin let an old nemesis creep in when he wrote, "I think there can be little doubt that use in our domestic animals strengthens and enlarges certain parts, and disuse diminishes them; and that such modifications are inherited." (Lamarckism at its most basic.)

This vagueness and vacillation in Darwin's writings caused confusion among critics and supporters alike. The lack of a sound biological explanation for variation weakened Darwin's argument and hampered acceptance of natural selection.

Ernst Mayr wrote that "The Darwin of 1859 was a pioneer, forced onto the role of an iconoclast. He would bravely state a new heresy on one page, then lose courage, begin hedging, and on a later page almost withdraw it altogether."

Like Wegener in his lifelong search for the solution to continental drift, Darwin missed the convergence of theory and technology in his quest for the source of variation. While Darwin was drafting the principles of evolution, an unknown Austrian monk and mathematician named Gregor Mendel was studying variation in pea plants. Mendel bred his plants through many generations and observed that characteristics such as flower color, stem height, and pod size would appear in statistically predictable patterns. His calculations suggested that traits are passed from generation to generation in the form of discrete units. Today we call these units "genes" and Mendel is known as the "father of modern genetics." Ironically, Darwin owned a copy of Mendel's 1866 book detailing the fundamental laws of genetic inheritance, but had never read it.

The knowledge that genes are the source of variation remained a mystery until rediscovery of Mendel's work in 1900. Today we know that surplus genes, small mistakes during gene duplication, and sex are just a few causes of variation.

Small changes in genes can yield big advantages. For example, hemoglobin is the complex protein in red blood cells that grabs oxygen from our lungs and delivers it to tissues and organs. A single chemical change in the gene coding for hemoglobin can protect red cells from being infected by the malaria parasite. This mutation can mean the difference between life and death for people living in areas where they are constantly exposed to malaria. But along with the good can come the bad. Two copies of the flawed hemoglobin gene, one from Mom and one from Dad, causes severe distortion of the hemoglobin molecule. Red blood cells become curved, carry less oxygen, and tend to clump together, clogging arteries and organs. The result is sickle-cell anemia, a disease that strikes down the young of African descent.

Sexual reproduction mixes and matches the variation found in our genes. Over eight million possible combinations can result from the sorting effect that occurs when our cellular machinery makes egg and sperm cells. After fertilization the resulting embryo contains a combination of genes arranged on chromosomes in an order that has never occurred before—and never will occur again.

What jump-starts **Speciation**—the generation of new forms of life? Both Darwin and Wallace agreed that ecology plays a central role in the emergence of new species. Refugee populations of animals, isolated from the larger group by barriers such as mountains or bodies of water, will forge unique relationships with their new environment. Natural selection preserves beneficial traits, and reproductive isolation from the larger population will concentrate and package those traits into a very different type of animal—a new species.

Small differences in physiology or behavior can also hinder reproduction and foster new species. Unusual size and shape of sex organs, novel mating calls that attract only a handful of females, or fertility cycles out of sync with the larger group can prevent some individuals from breeding normally with the rest of their species. These individuals may form a small founder population that breeds exclusively within itself and becomes a new species.

New species frequently emerge on the fringes of a normal habitat where small environmental differences are magnified. Only a few individuals will be able to adapt and survive under the increased selective pressures. The survivors will tend to breed exclusively among themselves, diverge genetically from the larger group, and eventually become a separate species.

Human enterprise on the landscape encourages environmental change and the emergence of new species. But it can also rapidly destroy and reinvent local environments, setting off alarms that call for drastic measures to protect a growing list of endangered species. For example, the northern spotted owl became a cause celebre in the early 1990s when extensive clear cutting in the Willamette National Forest of Oregon destroyed many old-growth pines that were its favored nesting sites. The number of breeding pairs dwindled and the owl was put on the endangered species list. As a result, federal and state timber lands were closed to logging and other ventures that might significantly degrade the owl's habitat. This move dramatically affected the local economy where residents depended largely on commercial lumber interests for their livelihood. A further move to prohibit logging on privately owned lands prompted legal action by landowners. The Forest Service and other government agencies have stepped in to shore up the sagging economy. Meanwhile, the resilient little owl has been seen nesting in new habitats further south and breeding with its cousin, the California spotted owl.

Extinction plays a crucial role in evolution. All species must change and adapt, or yield to those with a better design. Worldwide environmental disasters have not only annihilated species but also driven the emergence of many new ones, including humans.

Darwin said evolution takes place in incremental steps at a measured pace—the opposite of sudden creation. **Gradualism** became the fire wall in Darwin's theory when he wrote,

> *I do believe that natural selection will always act very slowly, often only at long intervals of time, and generally on only a very few of the inhabitants of the same region at the same time. As natural selection acts solely by accumulating slight, successive, favourable variations, it can produce no great or sudden modification . . . Hence the canon of 'Natura non facit saltum' [nature makes no jumps], which every fresh addition to our knowledge tends to make more strictly correct.*

Lyell had said the Earth transforms gradually over time, and it was a tempo that Darwin eagerly repeated. Darwin knew there were gaps in the fossil record, and that some species had existed for millions of years without changing. But he bet on the fossil record to prove a gradual tempo. As we'll see, Darwin's gamble became evolution's biggest nightmare.

Then and now many people misinterpret evolution as **progress toward perfection**. The *Origin of Species* is littered with statements saying exactly that. For example, "as natural selection works solely by and for the good of each being, all corporeal and mental endowments will tend to progress toward perfection." This narcissistic view tends to justify human superiority and makes evolution palatable for many. But careful reading reveals that Darwin saw perfection only in evolution's hindsight.

Natural selection operates within a perpetual flux of circumstances. Like middle-aged women who opt for facelifts, life is continually remodeling to ward off obsolescence. The winning design doesn't have to be perfect, just slightly better than the rest. Some designs appear to make no sense at all. For example, the magnificent tail of the male peacock or the lumbering weight of a bull elephant seal seem more of a hindrance than a help to the animal's survival. But the females of the species find these features irresistible. Sexual choice ensures an animal's reproductive and evolutionary success. The flexibility to change is essential. Perfection, if it exists, is fleeting.

Darwin also said evolution begins with *you*. This emphasis on the **importance of the individual** in the process of evolution may have stemmed from Darwin's interest in the laissez-faire economic theories of Adam Smith. Smith wrote that individual competition in a free market ensures a healthy economy. Successful people maintain their competitive edge by honing their singular skills. This leads to innovation and a more productive economy. In much the same way, individuals of a species allowed to compete unrestrained in the environment will make the most of their unique abilities. If they survive they will pass on their exceptional talents to future generations.

This scenario played out in ice-age Europe one-hundred-thousand years ago when a warming climate created new environments and set the stage for an encounter between the cold-adapted Neandertals and modern humans. When these small populations clashed, the value of individual survival skills were magnified.

But a new and controversial view of evolution's dynamics suggests that natural selection also operates beyond the level of the individual. Entire groups of species may be especially vulnerable to the pressures of natural selection.

Out of Africa: The Descent of Man

It's hard to imagine in today's dog-eat-dog world that a genuinely supportive relationship developed between Darwin and Wallace. But throughout his life Darwin was generous in word and deed to Wallace, using his contacts to secure a government pension for his impoverished friend. Wallace remained the humble admirer, often commenting that he was glad it was Darwin, and not he, who had undertaken the task of giving the world their theory. Wallace's role in developing the theory of evolution was well known among his peers and he was admired and respected until his death in 1913.

Wallace fully embraced natural selection as the only agent of evolutionary change. In that respect Wallace out-Darwined Darwin. But Wallace broke with natural selection when it came to human origins. As a super selectionist, Wallace insisted that natural selection must closely match structure and function to immediate need. But he knew that the brain of culturally primitive people was the same size and form as the brain of civilized Europeans. To Wallace this could only mean that the primitive's brain had been overdesigned. How could this have happened? Wallace used his belief in spiritualism to rationalize away this anomaly of natural selection. He rejected the idea that human intelligence and moral sense are a product of regular evolution. Instead he claimed that a form of spiritual evolution had produced the human brain for the sole purpose of guiding human consciousness and morality, just as natural selection had guided the physical evolution of the rest of the body.

In contrast, Darwin was certain that physical evolution could explain every aspect of becoming human. He tested the waters for this idea in the *Origin of Species*, writing that "light will be thrown on the origin of man and his history." Then in his 1871 book, *The Descent of Man*, he applied the same reasoning to human origins that he had applied to the evolution of fish, finches, and dogs.

His body is constructed on the same plan as that of other mammals. He passes through the same phases of embryological development. He retains many rudimentary and useless structures, which no doubt were once serviceable.

With cool logic Darwin pointed to the birthplace of humanity:

In each great region of the world the living mammals are closely related to the extinct species of the same region. It is therefore probable that Africa was formerly inhabited by extinct apes closely allied to the gorilla and chimpanzee; and as these two species are now man's nearest allies, it is somewhat more probable that our early progenitors lived on the African continent than elsewhere.

Darwin originated the notion of a "missing link" by associating humans with Old World monkeys and apes because of resemblances in our teeth and nostrils. He speculated that we share a common ancestor with the human look alikes, the gorillas and chimps, because we lack a tail. (see Figure 4.4)

Fig. 4.4 Model of Australopithecus africanus, a 2.5 million-year-old pre-human that walked upright on two legs. Could this creature be Darwin's missing link? (Copyright The Natural History Museum, London)

Darwin's simple, even crude logic represented a bold departure from scientific tradition. Belief in divine genealogy ran deep, forcing many scientists to bend the rules of classification. Some divided the living world into three kingdoms—Animal, Vegetable, and Human. More sensible minds placed humans among the animals, but not really one of them. Famed anatomist Richard Owen put humans in a separate subclass of mammals because of our brain, thus removing us from the taint of the monkey's tail. Thomas Huxley, Darwin's staunch supporter, agreed that humans are primates, but demurred by pigeonholing us in an order far removed from monkeys and apes. Darwin alone cut to the chase. He said if it looks like a monkey and acts like an ape, the rest is evolution.

The evidence from modern anthropology and genetics confirms Darwin's insight. Today humans are grouped with gorillas and chimpanzees. Less than a 2% difference in our genes separates us from the chimps—our closest living evolutionary ancestor. This minuscule degree of chemical separation belies major differences in expression of a common genetic heritage.

What does the phrase "missing link" suggest to you? For most of us it's an ape-like creature standing on the evolutionary threshold of humanity. We are curious about the finishing touches that made us fully human.

> Jack M., a young fifty-something, has been in the luxury car business for over thirty years. But on vacation he shuns lavish hotels and personal amenities to live in a tent and dig in the dirt at fossil sites around the world. He's found out a lot about human evolution, but not the answer to his biggest question.
>
> *To me it's interesting that as we learn more through anthropology and archaeology, we've been able to find humanoid types that go back beyond the first appearance of Homo sapiens—modern humans. You can see how the first of those creatures began stretching up to reach food, and eventually became bipeds instead of quadrupeds. Development seems to have stagnated in the apes, though chimps can make tools.*

> *But in the primate group that became humans there was further development. Those early creatures were very much like us. Even Neanderthals buried their dead like modern humans. But they weren't us. I cannot see a reason for the change to modern humans. There is a gap in evolution between the ones that are becoming less and less apelike, and Homo sapiens. I call it the missing link.*

Darwin countered the argument that the human mind is unique by pointing out that "every one who admits the principle of evolution, must see that the mental powers of the higher animals, which are the same in kind with those of man, though so different in degree, are capable of advancement." Our intellectual powers evolved.

Darwin drew similar parallels between the instincts of social animals and the ethics of humans by claiming that "the first foundation or origin of the moral sense lies in the social instincts, including sympathy; and these instincts no doubt were primarily gained, as in the case of the lower animals, through natural selection."

Darwin envisioned the human animal descending through time from simpler life, not descending from the divine. His early writings captured his uncensored opinion on the origin of the soul: "Plato says in Phaedo that our 'imaginary ideas' arise from the preexistence of the soul . . . read monkeys for preexistence."

Devout Victorians complained that Darwin had robbed them of the poetry of human existence. In 1864 Benjamin Disraeli, who was to become British prime minister, proclaimed to the Oxford Diocesan Conference, "Is man an ape or an angel? I am on the side of the angels. I repudiate with indignation and abhorrence those new fangled theories."

Was Darwin a believer or an atheist? Publicly a good old boy, privately a rebel, Darwin endured the conflict between his mind and his milieu. He displayed a public sensitivity toward the Christian beliefs of his society and audience. But the illicit, boldly materialistic tone of his private notebooks reveals an excited visionary entertaining profane thoughts. Darwin wrote to his close friend, the botanist Joseph Hooker, that to voice his belief in evolution and natural selection was "like confessing a murder."

Darwin's *Origin of Species*, the bible of modern day evolutionists, has as much to do with how we view the natural world and our place in it, as it does with the development of life. It also reveals the problem Darwin encountered when describing a power of nature that exceeds human experience. Language can only get so close to the truth. Unfortunately Darwin chose the familiar analogy of domestic breeding to illustrate the power of natural selection. Prolific use of verbs like "improving" and "selecting" seemed to imply intelligent choice. Some scholars have interpreted this to mean Darwin was a deist who believed in a divine intelligence behind evolution. The final passage of his book contains enough ambiguity to satisfy almost anyone.

> *It is interesting to contemplate an entangled bank, clothed with many plants of many kinds, with birds singing on the bushes . . . and to reflect that these elaborately constructed forms, so different from each other, and dependent upon each other in so complex a manner, have all been produced by laws acting around us.*

> *There is grandeur in this view of life, having been originally breathed into a few forms or into one; and that . . . from so simple a beginning endless forms most beautiful and most wonderful have been, and are being evolved.*

But Darwin reveals his gradual loss of faith through his letters and an extraordinary autobiography intended only for his wife and children. He boarded the *Beagle* as an orthodox believer, and disembarked a scriptural skeptic. During the voyage he often experienced a sense of the spiritual, writing in his Journal that when standing in the middle of a Brazilian rain forest, "it is not possible to give an adequate idea of the higher feelings of wonder, admiration, and devotion which fill and elevate the mind." But disenchantment with Christianity followed shortly after his return.

> *I had gradually come, by this time [1836 to 1839], to see that the Old Testament was no more to be trusted than the sacred books of*

> *the Hindus, or the beliefs of any barbarian. By further reflecting . . . that the more we know of the fixed laws of nature the more incredible do miracles become, I gradually came to disbelieve in Christianity as a divine revelation.*

After the death of his ten-year-old daughter Annie in 1851, Darwin lost any trace of his earlier religious leanings.

> *I had no intention to write atheistically. But I own that I cannot see as plainly as others do . . . evidence of design and beneficence on all sides of us. On the other hand, I cannot anyhow be contented to view this wonderful universe, and especially the nature of man, and to conclude that everything is the result of brute force (1860).*

> *In my most extreme fluctuations I have never been an Atheist in the sense of denying the existence of a God. I think that generally (and more and more as I grow older), but not always, that an Agnostic would be the more correct description of my state of mind (1873).*

Evolution's Smoking Guns

It's one thing to conceive of something, but another to actually see it happen.

Chances are less than zero that during our short lifetime we will be present the moment a new species emerges. Evolutionary time is measured in millions of years. Change can be a long, slow process. Even bursts of evolution take place on a time scale of tens of thousands of years. More to the point, it is next to impossible for humans contemplating a maximum lifetime of a hundred years to imagine a million years' worth of time, much less a billion. We can only fathom fragments of the process of evolution and project them through geological time.

One of those observable fragments is *microevolution*. Small changes in animals can offer survival advantages, but don't necessarily result in new forms of life. Probably the most famous example of microevolution is the English peppered moth, a species that exists in light and dark color variations. During daylight the moths rest on lichen covered trees that provide camouflage for the paler moths. Darker moths stand out like beacons for predators. Collections from the 1840s of peppered moths that lived near Birmingham, England reveal that the darker moths constituted only 1% of the total population. By 1900 sooty plumes from coal-fired factories of the Industrial Revolution had blackened the English countryside, killing off the lichens and darkening tree trunks. Dark peppered moths soon made up over 95% of the population. Films from the era show birds selectively homing in on the paler moths. In the 1950s Britain passed air pollution laws that reduced the soot. Lichens returned and, as Darwin would have predicted, the pale peppered moths quickly made a comeback.

Excellent example of evolutionary adaptation, survival, and flexibility? Absolutely. Evidence for major adaptive change and new kinds of animals? No.

But today a rise in popularity of field research is yielding evidence for real time evolutionary changes associated with the formation of new species. Studies of Australian fruit flies and wild sockeye salmon in the Pacific Northwest have confirmed that factors important to the emergence of new species can evolve rapidly in response to environmental change. For example, volatile hormonal signals called pheromones that identify individuals of a species and attract mates can become modified in fruit flies in the space of a few generations. The body shape of salmon will become dramatically altered in fifty years or less in response to changes in the landscape of their spawning grounds. These important cues for sexual selection and reproductive isolation were once thought to be passive by-products of more significant changes. They now appear to be the direct result of the selective pressures of evolution, operating on small changes within a time scale that is accessible to experiment and direct observation.

Darwin's greatest challenge was to explain how small changes could build complex organs like the eye or the brain. He wrote in the *Origin of Species*, "If numerous gradations from a perfect and complex eye to one very imperfect and simple, each grade being useful to its possessor, can be shown to exist, . . . then the difficulty of believing that a perfect and complex eye could be formed by natural selection, though insuperable by our imagination, can hardly be considered real."

Darwin's theory predicts that a small spot of light-sensitive pigment, providing a slight but important advantage in sensing movement, could have evolved independently many times. But creationists argue that eyespots are a far cry from the complex eyes of insects or humans.

Today exciting new discoveries take us back in time to the common origin of an indispensable building block of sight. The amazing extremophiles are members of the oldest and most primitive group of cellular life. These bacteria-like organisms live life on the edge in boiling temperatures of deep ocean vents and extremely acidic waters. Some of them have a distinctive red color that comes from a light-sensitive pigment that supplies them with chemical energy. This ancient molecule is chemically similar to rhodopsin, the vital light detecting molecule found in the retina of all vertebrates including humans.

Differences in design and performance of the eyes of many creatures have cried out for separate evolutionary pathways. For example, the multi-faceted compound eyes of insects are fashioned from hundreds of individual lenses arranged like tiny cameras. They can focus on the big picture and detect the slightest movement, but can't take the close-up, detailed shot. On the other hand, the eyes of worms, mice, and humans share a very different design. A single lens focuses light onto a patch of receptor cells that in turn stimulate nerves to produce an image in the brain. Within the receptors are thousands of proteins that absorb energy and enable most vertebrates to see details even in dim light.

Now genetic research is unraveling the close genetic relationship among all eyes. A gene known as *pax6* appears to be the universal instruction manual for making eyes. It has been found in nearly every animal, including vertebrates and insects. This master gene, and its role in eye development, suggest that dissimilar complex organs like the eyes of flies and humans can share the same evolutionary pathway. Natural selection started out with the same genetic assets, the same hammer and nails, to construct eyes in all their magnificent variety.

Theories of science, unlike doctrines of faith, have predictive value. As technology becomes available to test those predictions, the details of a theory emerge. Einstein's theory of relativity is tested everyday by new discoveries about the intimate workings of our Universe. These finds have confirmed the soundness of his theory. The same is true for evolution.

For example, there are two kinds of whales, those with teeth and those that have baleen to strain food out of water. Evolution predicts that there was a common ancestor of whales that had both teeth and baleen. That fossil ancestor has been found. Abundant fossil and genetic evidence attests to a course of evolution predicted by Darwin's theory. Scientists are still working out the details.

Margaret B. is a retired office manager in her seventies who shares an interest in science with her husband. She is a self-described atheist and holds no romantic notions about life's development.

It would not be possible for me to believe in both evolution and divine intervention. I think most scientists believe in evolution. They have to clearly separate their personal beliefs from their work. Perhaps the more rigorous the science the more likely they are to take a cold, analytical approach to the question of the origin of life.

I have known a lot of astronomers who are atheists, and some chemists and M.D.'s who were believing people. But my husband and I went to an astronomy convention once, and we were surprised to see there were a lot of nuns in the audience. I guess they were teachers.

National surveys show that one out of ten Americans would agree with Margaret, a strict Evolutionist. There is no room for doubt or the divine in their belief that life evolved solely by means of natural laws.

A much larger number of my Survey participants (60%) agree that the theory of evolution correctly explains human origins and the development of life on Earth. They also overwhelmingly believe (74%) that most scientists accept the modern theory of evolution as valid. Surprisingly nearly 42% of these same folks reject the notion that all scientists are atheists. And in fact, national surveys have also demonstrated that nearly 40% of American scientists believe in both evolution and a spiritual cause for human origins.[1]

The Great Darwin Divide

When Thomas Huxley first heard of Darwin's theory, he exclaimed, "How extremely stupid of me not to have thought of that."

Darwin's simple scheme freed natural science from its twenty-five-hundred year marriage to Plato's ideal forms. If disenfranchising fixed species wasn't enough, Darwin's emphasis on natural selection's unlimited creativity sealed their fate.

But no sooner had Victorians learned that human evolution is a natural biological process than they tried to control it. Darwin's cousin Francis Galton, a leading eugenicist, complained in his book *Hereditary Genius* (1869) that highly intelligent people weren't having enough children, and the human race would decline rapidly if the trend were not reversed. Aldous Huxley, grandson of Darwin's advocate Thomas Huxley, wrote the perennial cult classic *Brave New World* (1932) to warn of a future Utopia in which rules based on natural selection would govern society, and replace free will with controlled reproduction.

In light of today's science, many of Darwin's speculations in the *Origin of Species* appear prophetic:

- *A grand and almost untrodden field of inquiry will be opened on the causes and laws of variation* (The sciences of Genetics and Molecular Biology)

- *The study of domestic production will rise immensely in value. A new variety raised by man will be a far more important and interesting subject for study than one more species added to the infinitude of already recorded species.* (Genetic Engineering of crops and cloning of livestock)

- *Embryology will reveal to us the structure of the prototypes of each great class.* (Developmental Genetics reveal the evolution of complex structures)

- *In the distant future I see open fields for far more important researches. Psychology will be based on a new foundation, that of the necessary acquirement of each mental power by gradation.* (Animal Psychology and the evolution of consciousness)

During the 1940s geneticists and field biologists reached a consensus known as the Modern Evolutionary Synthesis. They endorsed Darwin's theory of natural selection and agreed that variation, the raw material of evolution, is in our genes. This scientific marriage was consummated in the 1950s when James Watson and Francis Crick unravelled the double-helix structure of DNA and its function as the template for gene duplication and inheritance.

If Darwin was right, why does his theory remain so controversial? It was only as an exercise in "what if" that Darwin extended his theory to encompass the origin of all life when he wrote, "Probably all the organic beings which have ever lived on this earth have descended from some one primordial form into which life was first breathed."

The self-professed agnostic routinely separated his religious views from his science. But in a great leap of (scientific) faith, Darwin moved from the observable and palatable to the truly visionary. It was a tiny step for those able to take it. Darwin took it, and that is at the heart of the divide.

It is difficult, if not impossible, to champion two conflicting philosophies. Darwin's theory left little room for mystics. Richard Dawkins, evolutionist and staunch Darwinist, has said that Darwin gave us the answer to *why* we exist—a question usually reserved for religion. The late Stephen Jay Gould, a pioneering "new evolutionist," placed Darwin's theory and traditional theology on equal but separate footing because they give us the answers to very different questions. Ernst Mayr says Darwin's theory enables the biological sciences to answer both proximate (how) and ultimate (why) questions, but perhaps not in a spiritual sense.

Today zealous creationists vilify Darwin and claim the weight of scientific proof for biblical myth. Their answer to the Great Darwin Divide goes beyond Genesis to a takeover attempt targeting our schools' moral and mental yardsticks.

Evolution is like a three dimensional puzzle. Its pieces and players move in many directions on multiple levels at different speeds. Their interactions are often elusive, and many questions remain. Next we'll look at evolution through the eyes of the detectives who probe evolution's process, and uncover the truth behind evolution's ugly little facts.

NOTES

1. Edward J. Larson and Larry Witham, "Scientists are Still Keeping the Faith," *Nature*, Vol. 386, April 3, 1997, pp. 435–436; George Bishop, "The Religious Worldview and American Beliefs About Human Origins," *The Public Perspective*, Vol. 9, No. 5, The Roper Center, University of Connecticut, 1998, pp. 39–44.

REFERENCES

Barlow, Nora (ed.). 1958. *The Autobiography of Charles Darwin, 1809–1882*. New York: W.W. Norton.

Burkhardt, Frederick (ed.). 1996. *Charles Darwin's Letters: A Selection 1825–1859*. Cambridge: Cambridge University Press.

Darwin, Charles. 1859. *On the Origin of Species.* London: John Murray.
———. 1871. *The Descent of Man.* London: John Murray.
Darwin, Francis (ed.). 1958. *The Autobiography of Charles Darwin and Selected Letters.* New York: Dover.
de Beer, Gavin (ed.). 1979. *Charles Darwin and T. H. Huxley: Autobiographies.* London: Oxford University Press.
Gould, Stephen Jay. 1977. *Ever Since Darwin.* New York: W.W. Norton.
Huxley, Julian. 1958. "Introduction to the Mentor Edition." *On the Origin of Species.* New York: The New American Library.
Mayr, Ernst. 1964. Introduction to the facsimile edition of *On the Origin of Species.* Cambridge: Harvard University Press.
———. 1991. *One Long Argument.* Cambridge: Harvard University Press.

WEBSITES

BBC Evolution Website (http://www.bbc.co.uk./education/darwin) Packed with information on Darwin, the evolution debate, articles by leading scientists and clergy, and links to other evolution and fossil websites.

The Alfred Russel Wallace Page (http://www.wku.edu/%7Esmithch/home.htm) Fascinating, well researched insight into the life and work of the English naturalist, social critic, and co-discoverer of the modern theory of evolution.

SUGGESTED READING

Darwin's Ghost: The Origin of Species Updated, Steve Jones, Random House, 1999.
A retelling of Darwin's classic work with contemporary examples, latest research, anecdotes, and humor. Jones has been called "the British Carl Sagan" because of his fame as a popularizer of science.

Chapter 5

Ugly Little Facts

The great tragedy of Science— the slaying of a beautiful hypothesis by an ugly fact.

Thomas Henry Huxley, Presidential Address to the British Association for the Advancement of Science, 1870

Evolution isn't what we always thought it was.

Dorothy F. is a retired foreign service officer in her sixties. Though raised in a conservative Midwestern family, she soon began to look outside her religion for answers to life's biggest questions. Dorothy describes herself as "intellectually interested in everything about human evolution," and believes that she knows where we came from.

I believe, from what has been discovered in Africa, that we have gradually evolved into the human beings we know today—Homo sapiens. That's it. I think we evolved naturally. I do not believe, quite frankly, in an all-seeing Being that created Earth as it is.

I came to my beliefs gradually. As a teenager, I asked questions. In college I studied how religions evolved as part of man's desire for structure. I do not need a structure, such as a metaphysical Being, to lean on. So belief in natural evolution follows.

I think the so-called mysteries exist only in those things that haven't been discovered. I don't think there's any such thing as the mystery of life or the mystery of the Universe or the mystery of the evolution of man. I think all the answers are out there. It doesn't bother me that they haven't been found yet.

Evolution may not be Darwin's evolution as such. We may not have any natural selection or natural evolution as we think of it.

But I still believe that it is not part of an overall plan by an all-seeing Being greater than we are. I still believe it's a natural thing that happens and has happened, and it's just a question of not knowing enough about it yet to be specific. Why did life evolve on the Earth? It just did. The environment was right.

Dorothy is typical of the 10% of Americans who believe that natural processes alone were responsible for human origins and the beginning of life. In Dorothy's mind, evolution just happened somehow. But in that somehow lies the problem for others. My Survey and interviews reveal that many people embrace the principles of Darwin's theory, but don't buy into all its conclusions. They sense voids in our knowledge about the process of evolution and aren't totally convinced by the smoking gun evidence. These folks are buyers of evolution, but need to be sold on the steps. Perhaps you are one of them.

Jeannette K. is a sophisticated woman in her fifties whose charming accent reveals her European heritage. She is a serious traveler with an educated appreciation for the diversity of world cultures and religions. Jeannette spoke with us just before leaving on a trip to Easter Island.

I don't think I believe in the religious theory. I think it was an evolution, probably a chemical evolution of some sort, like the Big Bang. But I don't really accept the religious explanation of God creating the Universe. I have problems with that, but at the same time, maybe it's possible.

When I was a teenager I remember being bored at Mass. That was a time when I was studying the Reformation and thinking about the whole issue of religion. People can believe in evolution and in some supernatural power. I believe in the natural evolution suggested by science. But I don't believe it has to be the only sure thing.

Some things about evolution still bother me, questions that need to be answered. For example, why did the dinosaurs die? Do we know the explanation? I mean for sure? I know there have been climate and other things mentioned, but do we know 100% for sure?

What is it about evolution that we find hard to accept? For one thing, we've never really seen it occur. In our short lifetimes most of us will never witness the emergence of a new species. We have seen compelling evidence that random variation and natural selection can bring about changes in organisms that can lead to new species. But most of us think of evolution as change on a grand scale. Science calls this *macroevolution*—major genetic change that spawns entirely new kinds of animals. There has never been an observed and documented instance of macroevolution.

Furthermore, the evolutionary history of many animals remains unclear. New finds often reveal that ancient species didn't vanish but continued to live alongside newer species for considerable periods of time. Evolutionary trees have become evolutionary bushes. Perhaps the best example of this phenomenon can be found in the origin of modern humans.

The first Neandertal skeleton was found three years before Darwin published his theory. Since then controversy has swirled around the place of Neandertals on our evolutionary family tree. Scientists generally agree that we share a common ancestor with Neandertals called *Homo erectus*, a pre-modern human species that migrated out of Africa into Europe and Asia nearly two million years ago. Neandertals are the descendants of *Homo erectus* in Europe, but appear to have vanished around the time that modern humans migrated from the middle East into Western Europe.

Conventional wisdom has long held that Neandertals were overwhelmed by competition from more technologically advanced humans, and became a dead-end branch. But new evidence suggests that Neandertals and modern humans coexisted in locations from Israel to Spain for thousands of years. Could they have lived in close proximity for long periods of time without sharing their genes as well as their tools?

Charles Darwin gave science its most powerful weapon. No single scientific theory has been as influential in shaping social and scientific philosophy. Phrases like "natural selection" and "survival of the fittest" have become part of our vocabulary. Nearly a century and a half of research and the advent of sophisticated technologies have helped us decode the evidence for evolution. Yet nagging questions remain.

The Problem with Fossils

Darwin gambled on the fossil record to prove the gradual tempo of evolution. But paleontologists had long known that there were gaps in the fossil trail. Darwin insisted that the gaps were simply an "imperfection of the record" due to rarity of fossilization, poor preservation, and insufficient exploration of fossil sites. He was convinced that as more fossils were found the gaps would close. Close friends like Thomas Huxley warned Darwin, "You have loaded yourself with an unnecessary difficulty in adopting 'Natura non facit saltum' (nature makes no jumps) so unreservedly."

The gaps persisted, prompting Darwin to write, "Why does not every collection of fossil remains afford plain evidence of the gradation and mutation of the forms of life? We meet with no such evidence, and this is the most obvious and forcible of the many objections which may be urged against my theory. We are always slow in admitting any great change of which we do not see the intermediate steps."

Since Darwin's day, the fossil record has grown exponentially. Rocks from all periods of geological time have yielded up their secrets to the paleontologist's pick. Literally hundreds of thousands of new fossil species have been added to our picture of evolution. Over one hundred new fossil sites worldwide containing large numbers of rare fossils provide snapshots of the rich variety of life in ancient environments. One of these sites is the Burgess Shale, a fossil-rich outcrop in the Canadian Rockies near Banff that reveals an amazing explosion of multicellular life during the Cambrian period half a billion years ago. The Cambrian was once viewed as an age of evolutionary experimentation, but is now seen as the time when major animal groups such as vertebrates first appeared.

If we could play back the videotape of evolution revealed by these spectacular fossil finds, we would see patterns very different from those predicted by Darwin. New kinds of animals seem to appear suddenly in the fossil record. Evolutionary trails of complex organs and unique structures are difficult to trace. The more fossils found, the more obvious the gaps. Fossil gaps remain paleontology's dirty little secret.

More Ugly Little Facts

June through November is hurricane season in Florida. Tropical storms, heavy surf, and extreme tides pound the shore, leaving behind a chronicle of life's past. Marooned sea creatures, so perfectly shaped that they look artificial, remind us of a time when prehistoric seas washed across North America. Ancient ancestors of these animals were once among the most abundant life forms on Earth.

Perhaps the most primitive looking of all are the horseshoe crabs. Armed with reddish-brown shields and menacing spikes, they look like miniature soldiers from an alien world (Figure 5.1).

Fig. 5.1 Horseshoe crabs are called living fossils because they have evolved at such a slow rate that their appearance is virtually unchanged from that of their ancestors two-hundred-thirty million years ago.

Horseshoes are not really crabs at all, but distant relatives of spiders. They are often called *living fossils*—a term Darwin coined to describe animals he thought were relics from the past that had stopped evolving. In fact the living species of horseshoe crab can trace its evolutionary history back at least two-hundred-thirty million years. But changes in horseshoe crabs during that time have been so slight, and new species so few, that today's versions are virtually identical to horseshoes that swam in Jurassic seas when dinosaurs ruled the land. The horseshoe isn't stuck in an evolutionary backwater, but simply represents a group of animals that has evolved at an extremely slow rate. Contrary to Darwin's theory, horseshoe crabs evolve at a different rhythm from that of the steady, gradual change leading relentlessly to new species.

Let's contrast the horseshoe crab with its distant relative, the spider. Spiders are evolution's champs. They have flourished for over three-hundred-million years, radiating into diverse habitats ranging from sea level intertidal flats to the windswept Himalayas. They are found on all continents except Antarctica. The total number of spider species outnumber all species of fish, amphibians, reptiles, birds, and mammals combined.

Modern spiders fall into two major groups. The larger group contains nearly 35,000 species, including common spiders and tarantulas. But a second smaller group consists of only 40 species whose habitats are restricted to China, Japan and Southeast Asia. These Oriental spiders have primitive features that set them apart from more evolved species. Many are virtually indistinguishable from three-hundred-million-year-old fossil spiders. Oriental spiders are another example of the unevenness of evolution's tempo, seen here as a small, slowly evolving branch of an otherwise rapidly evolving, highly diversified group of animals.

There are two special criteria that living fossils like horseshoe crabs and Oriental spiders must meet:

1. They must belong to a distinctly evolving animal group that exhibits a very low rate of species formation. Often there may be only a single living species.

2. The modern species of living fossils must retain specific features that link them to their ancient ancestors. Living fossils simply look old.

Examples of living fossils include the American alligator, porcupines, snapping turtles, and the Coelacanth—an enormous fish considered extinct for the last eighty million years (Figure 5.2). Coelacanths were rediscovered in 1938 in waters off the coast of Africa. In 1998 a second living species was found six-thousand miles away near Indonesia, hinting at more populations of this living fossil.

Fig. 5.2 Coelacanths bear live young and have fins that move like limbs of land animals. They can reach lengths up to six feet and weigh over 200 pounds.

Living fossils posed a problem for Darwin because they didn't conform to the predicted pattern of constant change and diversity. He suggested that living fossils "have endured to the present day, from having inhabited a confined area, and from having thus been exposed to less severe competition." But many living fossils are widely adapted, broadly distributed, and extremely abundant, as was the American alligator before being hunted to near extinction in the early part of the 20th century. Furthermore, living fossils are all very different. They don't appear to share a common trait that might predispose them toward a slow rate of evolution.

Slow rates of evolution spell success for some species, and faster rates of change and diversity become key to the survival of others. Fossil gaps. Living fossils. Was Darwin right or wrong? The answer is—not exactly. Darwin got it mostly right. Let's see what the controversy is all about.

The ABCs of the Evolution Controversy

Public relations has never been science's strong suit. Announcement of a significant discovery is followed by years of silence. Anthropologists and paleontologists prefer to return to the field where each day brings the possibility of unearthing the oldest or the newest—their version of winning the lottery. Descriptions and interpretations of new evidence become buried in the unintelligible jargon of scholarly journals. The resulting information blackout reinforces the notion that evolution is a tentative theory supported by sketchy evidence.

Scientists further weaken our confidence when they debate the precise mechanism of evolution. Their discussions are highly theoretical, confusing to all but the initiated, and open to as many degrees of interpretation as there are evolutionists. It may seem to many of us that even the evolutionists aren't sure about evolution.

Incomplete and contradictory information continues to reinforce diverse concepts of evolution. For example, many people characterize evolution as simply the genetic relationship between humans and apes. Others believe evolution involves a purposeful, steady progress from lower to higher forms even though Darwin consistently argued that evolution has no purpose or direction. The natural processes that play a role in evolution have also become part of its definition. For example, we often portray evolution as meaning "the strong eliminated the weak" or "some organisms left more offspring than others."

Facts about evolution often fall victim to the public's fascination with things ancient and alien. A survey conducted by the American Museum of Natural History revealed that over 35% of Americans believe dinosaurs and humans lived on the Earth at the same time. In fact, at least sixty million years separate the last remnants of the dinosaurs from the appearance of the earliest known human ancestors. Such misconceptions can be unwittingly fueled by the most innocent of circumstances. An educational gift shop at Florida's popular Disney World displays miniature plastic models of dinosaurs alongside small figures of prehuman primates walking upright and cradling children in their arms. Anyone looking at this arrangement could assume that all those creatures lived at the same time.

Popular science programs like *NOVA*, and magazines such as *Discover* and *National Geographic*, try to give readers accurate coverage of new discoveries. But their reports are often overshadowed by the sensationalism of the discovery itself. For example, "Large and Lungless" is the title of an April 1996 *Discover* article that claims an extinct amphibian, the largest known, may still survive in South America.

Often it's the philosophical debate surrounding evolution that gets star billing. A front-page story from *The Wall Street Journal*, titled "Darwinian Struggle: Instead of Evolution, A Textbook Proposes 'Intelligent Design,'" examines the controversy surrounding the teaching of a quasi-religious concept of evolution as a scientific alternative to Darwin's theory.

When fragmentary reports, confusing science, and pseudoscientific explanations run headlong into strongly held beliefs, they can result in amazing feats of imagination. Each of us has a personal impression of evolution that's been shaped by education, life experience, and religious belief. Science can clarify that picture and sell us on evolution by explaining the ABCs of the evolution controversy.

A—The Abracadabra Effect

Bats are the only mammals that fly. They are the premier night hunters of the sky, locating and capturing prey on the wing in total darkness. They literally whistle through their noses, sending out short bursts of ultrasonic sound waves that bounce off insects and moths. Return echoes are collected through the bat's oversized ears and its neurological system instantly reads them to determine location, speed, course, and size of the moving target. Evolution of the bat's complex survival skill called *echolocation* involved the co-development of multiple systems (vocal, neurological, auditory) and structures (larynx, ears, eyes).

But there are no pre-bats. All the features needed for sustained flight and echolocation were present and fully formed in the earliest known bats. *Icaronycteris* is one of the oldest fossil bats dating from fifty million years ago. It has all the structures of modern bats—including large ears to catch echolocation signals and muscular wings for true flight. Some scientists suggest that bats, lemurs, and primates have a common ancestor. But no intermediate forms of bats appear in the fossil record.

Abracadabra!

The true magnitude of the gap problem emerges if we try to estimate how much time and how many genetic changes are needed to produce a new kind of animal with distinct features like the bat. How would you tackle a problem like this?

Evolutionists start by sorting animals into categories based on their shared traits. These common characteristics can also suggest common descent (Figure 5.3).

For example, humans are in a Class of vertebrates or back-boned animals called Mammals. The Mammal class also includes elephants, wolves, dolphins, and bats, among others. All mammals share certain identifying characteristics, such as mammary glands and the bearing of live young. But the class is very large, and its members can also display extreme differences. So to further simplify things, mammals are grouped into Orders based on an

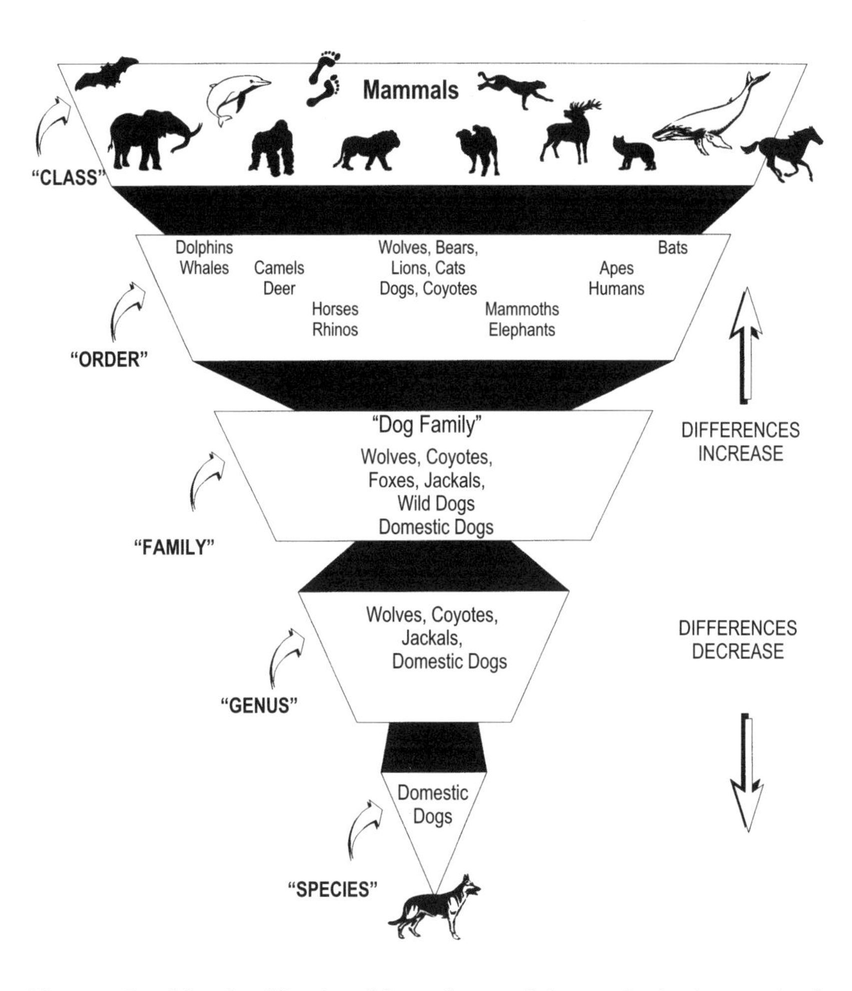

Fig. 5.3 In this classification hierarchy used by evolutionists, animals grouped in the same category at each level share similarities in physical traits. But the degree of difference among animals increases as we move up the hierarchy.

increased similarity in features. Wolves, bears, coyotes, lions, cats, and dogs are more similar to each other than they are to dolphins and whales, so they are grouped together in a separate order.

Further sorting creates the Family, Genus, and Species categories. For example, wolves, domestic dogs, wild dogs, and foxes are all placed in the same Family because they are more similar to each other than they are to bears or lions. But wolves, coyotes, jackals, and domestic dogs have been placed in their own Genus since their traits and inherited characteristics have shown them to be more similar to each other than they are to wild dogs and foxes. Finally, all domestic dogs are considered a single species because they interbreed easily and are virtually identical. Though centuries of artificial breeding have exaggerated certain traits to produce breeds like the German shepherd and the poodle, we know from fossil and genetic evidence that man's best friend is fundamentally a single species, most closely related to the wolf.

When we follow this classification system in reverse, we see that the magnitude of difference increases as we move up to the next category. Characteristics that define higher categories become more unique. For example, the degree of difference between a dog and a whale is far greater than the degree of difference between a wolf and a dog. Traits that separate animal groups at successively higher levels represent pivotal evolutionary innovations. What were the natural processes that led to these major evolutionary changes?

By the time the dinosaurs died out, primitive, shrew-sized mammals had been around for at least one-hundred-fifty million years. But during the ten to fifteen million years following extinction of the dinosaurs, a burst of evolutionary activity led to the emergence of most Orders of modern mammals. Among these were carnivores such as lions, wolves, and bears, as well as herbivores and omnivores such as elephants, horses, whales, rhinos, camels, pigs, deer, and bats. When these major animal groups first appear in the fossil record, they have already evolved the major innovations of form and function that distinguish them today.

How can small adaptive changes assemble so many unique forms of life in such a relatively short time period? Darwinists like Richard Dawkins argue that these spans of time in concert with the measureless creative power of natural selection are all that's needed. But consider this. Fossil dating reveals that the average life span of a mammal species from its first appearance to its apparent extinction is one to two million years. If Darwin was right and species gradually branch into new ones through incremental change, then the fifteen million years between the fall of the dinosaurs and the rise of modern mammals would permit the evolution of only fifteen new *species*—much less a dozen major groups.

Small genetic differences can, in fact, account for surprising variations in anatomy, physiology, and behavior in people. For example, a close childhood friend was the recipient of the only gene for carrot-red hair in three generations of his handsome, dark-haired family.

But larger genetic modifications, either naturally occurring or experimentally induced, often result in non-viable offspring or grotesque monsters. For example, a common experimental mutation in fruit flies produces living offspring with legs sprouting from their heads. If most large mutations are lethal or non-functional, what type of genetic change powers major evolutionary innovation?

Startling new discoveries have revealed a hidden source of genetic potential. Genes have been sorted, rearranged, and fine-tuned by the assault of millions of years of evolutionary pressure. The work of Christopher Wills and other molecular biologists suggests that there is an accumulated wisdom in this organization that enables genes to work more efficiently and promote innovative change.

For example, the arrangement and action of developmental genes that control the pattern of growth in an embryo are strikingly similar in dissimilar organisms. Most vertebrates, including mammals, have four similar gene clusters controlling growth of everything from arms to eyes. The same genes that promote limb and rib development in mouse and chick embryos are also active in snakes. These genes have enormous leverage to effect major change through slight modifications.

How did snakes lose their legs and gain an elongated body and their slithering style of movement? Michelangelo thought the loss had occurred in the Garden of Eden. But a team of embryologists and developmental biologists have found a more mundane answer. They discovered that slight alterations in the activity of genes controlling limb and trunk development in chicks and snakes can lead to a loss of legs. These same genes are responsible for rib patterning, and could play a role in sculpting a snakelike body. This clearly demonstrates that a small change in a key developmental gene can produce major evolutionary innovations.

B—Bats, Birds, and Brains

George Bernard Shaw called Darwin's theory "a chapter of accidents" that "seems simple, because you do not at first realize all that it involves. But when its whole significance dawns on you, your heart sinks into a heap of sand within you. There is a hideous fatalism about it, a ghastly and damnable reduction of beauty and intelligence, of strength and purpose, of honor and aspiration."

Shaw's dismay at finding no sense of purpose and meaning in evolution still resonates in beliefs about where we came from. Over 29% of my Survey participants say evolution is purposeful and moves in a direction from lower to higher forms. When we blend our need for purpose with concepts like progress, complexity, and perfection, we draw powerful conclusions that are not easily shaken.

Today this picture of evolution as purposeful and progressive is often championed in popular literature because it elevates human beings, late arrivals on the evolutionary scene, to the status of most complex and perfect. It's a notion that's hard to deny.

In his best-selling book, *The Road Less Traveled*, psychiatrist Scott Peck claims,

> *And so it goes, up the scale of evolution, a scale of increasing complexity and organization and differentiation, with man, who possesses an enormous cerebral cortex and extraordinarily complex behavior patterns, being, as far as we can tell, at the top.*

In fact, nothing could be further from the truth. If evolutionary success can be measured by breadth of range, biological diversity, and persistence through time, then we are still living in the Age of Bacteria, not the Age of Mankind. In the fossil record of nearly four billion years of life on Earth, bacteria and other single-celled organisms have dominated our planet. Bacteria are the most prolific and most successful forms of life, inhabiting all environments from the frigid poles to the hot sulfurous vents of the ocean depths. But most of us find it hard to dismiss the idea that we are at the top of the evolutionary ladder and meant to be there. The evidence for this belief is literally in our heads.

Our earliest known primate ancestors had a brain capacity of 400 cubic centimeters, only slightly larger than that of a chimpanzee. One cubic centimeter is about the size of a sugar cube. Since that time the human brain has more than tripled in size and now hovers around 1350 cubic centimeters. We are each born with far more brain cells than we'll ever need. As many as a hundred billion neurons in our brain communicate with each other through complex chemical connections. Sensory activities are not isolated to specific structures like the cerebral cortex, but are the result of complex and dynamic interactions among many areas of the brain. Recent studies reveal that the size of the human brain peaked about thirty thousand years ago, but its size and functionality has obviously given our species an evolutionary leg up. However, it remains difficult for many people to accept the natural development of such complexity in a chunk of tissue weighing less than three pounds.

Darwin knew that his theory "would be unsatisfactory until it could be shown how the innumerable species inhabiting this world have been modified, so as to acquire that perfection of structure and co-adaptation which most justly excites our admiration." Complex structure and adaptations in nature continue to amaze us. We find it difficult to visualize the natural processes that could have produced the human brain or a bird's wing.

True flight is known to have evolved independently three times among vertebrates—in the ancient flying reptiles known as pterosaurs, in the bat, and in birds. In 1861 stonecutters working in a quarry near Solnhofen, Germany uncovered the one-hundred-forty-million-year-old fossil of a pigeon-sized animal with teeth like a small dinosaur and wings covered with feathers like

Fig. 5.4 This drawing of a fossil of Archaeopteryx, the oldest known bird, features a long bony tail, wings ending in needle-sharp claws, and the asymmetrical feather structure identical to that of modern birds.

modern birds. Dubbed *Archaeopteryx* or "ancient bird," the find instantly ignited controversy over the origin of birds (Figure 5.4).

Naturalist Thomas Huxley, Darwin's champion, labeled *Archaeopteryx* a prime example of evolution in action, a transitional form between reptiles and birds. It remains the oldest fossil of a modern bird. When Huxley spied a small dinosaur fossil lying alongside Archaeopteryx, he saw a strong resemblance and concluded that dinosaurs and birds were related—an evolutionary connection still hotly debated today.

Flight also evolved independently in bats along a different, but equally effective, design. Bats don't have feathers. Their wings are elongated finger bones covered by a thin membrane of skin. A bat's wing isn't a simple gliding airfoil like that of the wing of a flying squirrel. Instead, a bat's wing has muscles that power controlled, sustained flight. Random variation played a key role in the design of the wings of bats and birds.

Throughout his life Darwin wrestled with the source of the variation needed for evolution of complex structures. He was an avid breeder of pigeons, and based much of his argument for evolution's creative power on the results of domestic breeding. But selective breeding has its limitations. It may be possible to breed bigger and faster pigeons, but it's not possible to breed a pigeon into a goat.

Moreover, when an animal is bred for maximum development of specific traits, its overall health can suffer. Consider the magnificent Newfoundland, a dog prized and bred for its size and strength (Figure 5.5). Breeding of its oversized body has placed enormous stress on its limbs and cardiovascular system and shortened its lifespan to less than ten years.

Darwin was never able to resolve the dichotomy between a limiting force in nature and the immense amount of variation required for complexity. Even after the Evolutionary Synthesis of the 1940s, when both naturalists and biologists accepted genetic mutation as the source of variation, questions remained concerning the ability of our genes alone to support complex change.

Fig. 5.5 Centuries of selective breeding of Newfoundlands like "Moose," shown here, has resulted in sacrifice of the animal's health for size and strength.

Recently an exciting piece of the puzzle surfaced when scientists discovered a latent capacity for variation in our genes. When cells divide, a surplus duplication of some genes makes more material available for mutation, and increases the possibility that a novel change will arise that enhances survival.

In addition, the rearrangement of genes during millions of years of evolution has placed unrelated genes in close proximity and organized them into clusters that act as units. We saw this in the similarity of groups of developmental genes among different animals. These clusters are the genetic building blocks that assemble complex structures. A small mutation in a single gene is not an isolated event, but produces a ripple effect down the line. Coordinated changes across gene clusters controlling multiple systems could build complex adaptations like echolocation in bats. Major evolutionary advances can take place quickly because genes are organized for change.

C—Chance and Catastrophe

Chance is the single most misunderstood factor in evolution.

Beth K. is a high school science teacher and single parent in her mid-forties. Her strict upbringing in the Bible-belt South led to a degree in religion, but her fascination with science and nature determined the course of her life's work. Beth embraces the theory of evolution as scientific fact, yet acknowledges that her understanding of nature's complexities has only strengthened her belief that a divine controller set evolution in motion.

I believe in the theory of evolution because I think the thing that causes change in creatures, be they flora or fauna, is a Divine Being. I don't think that all the complexity and the interrelationships that we see between plants and animals, the way they actually function, could take place on a random, haphazard basis. I think there has to be a control mechanism that has placed all of this into being.

I can remember very specifically teaching about the transport systems of plants and thinking that this can't have just happened by chance because some atoms got together. There has to be a plan to this. And when you look at the development of a child, two cells coming together and all the parts of a human being that come out of those two cells, I just can't believe that it randomly happened. Because we were chemical soup and the complexity of life has sprung from chemical soup, I think there has to be something in control. It just makes sense to me.

I think anything's possible. But the odds against life developing as we currently have on Earth seem to me to be so great that it would be almost impossible, not actually impossible, but theoretically impossible without an ultimate cause.

Beth is a stealth Creationist. She believes in a process of natural evolution conceived and guided by an unnatural force. Her retreat to a belief in the impossibility of life's complexity places her firmly among disciples of Intelligent Design—a brand of creationism without the God-word.

In his book *The Blind Watchmaker*, British evolutionist Richard Dawkins calls a belief such as Beth's the argument from incredulity—"I just can't believe" or "It seems impossible." He explains that chance alone plays a very small role in the creative process of evolution. For example, the outcome of a roll of the dice is truly random, assuming you're not playing with loaded dice. Every roll is independent of all past rolls. But each step in evolution is contingent on the outcome of on all previous steps. This builder effect of evolution leads to complex results in which chance alone has played a very small role.

We can't accurately predict the outcome of evolution because we can never know all the elements that will influence the result. But we can often understand in hindsight why a complex evolutionary event took place once we are aware of all the factors (fossil history, genetic potential, environmental constraints) that were in play at the time the event occurred. This type of historical prediction is exactly what evolutionists do. For example, we can't at this time predict when or if humans will evolve a third eye. Such an event could be explained accurately only in retrospect.

Many evolutionists believe that Darwin's view of the role of chance in evolution was really one of contingency. Chance plays a role contingent on other factors that present the opportunity. But what happens when a truly random event places overwhelming stress on the dynamics of evolution? Could chance catastrophe put unique opportunities into play and change the rules of the evolutionary game? Patterns of diversity and extinction in the fossil record disclose at least five such events. We call them mass extinctions (Figure 5.6).

Mass extinctions are global disasters affecting all ecosystems. Entire lines of highly evolved, adaptively successful organisms are eliminated. The effects of mass extinctions are so clearly written in the fossil record that geologists use them to define important periods in evolutionary history. In each of the Big Five at least 65% of all living species perished. The worst occurred two-hundred-fifty million years ago at the end of the Permian period. This global catastrophe was at least twice as severe as any others, wiping out over

THE BIG FIVE		
Period	*Time Frame*	*Life Forms Affected*
Ordovician	440 MYA	Trilobites
Late Devonian	365 MYA	Marine life (trilobites, corals, fish)
Permian	250 MYA	Marine life, reptiles, amphibians
Triassic	210 MYA	Mollusks, arthropods, land vertebrates
Cretaceous	65 MYA	Dinosaurs; marine and flying reptiles
(MYA = million years ago)		

Fig. 5.6 The five largest extinction events in our planet's history and the animal groups most affected.

95% of all living species. Marine life was devastated. Hardest hit were mollusks, corals, sea urchins, and ancestors of the modern squid and octopus. Trilobites disappeared forever. Extinction of over 70% of all higher animal species on land saw the loss of most groups of amphibians and reptiles.

The most recent mass extinction occurred sixty-five million years ago at the end of the Cretaceous. Over 50% of the world's species disappeared. Though far less severe than the Permian extinction, it is better known because it signalled the end of the long and successful reign of the dinosaurs.

In the early 1980s, Nobel Prize-winning physicist Luis Alvarez and his son Walter, a geologist, found high concentrations of a rare element called iridium in red clay layers dating to the time of the Cretaceous extinction. Iridium is seldom found in large quantities except in meteorites and some volcanic eruptions, so the pair developed a theory suggesting that sixty-five million years ago the Earth took a major hit from a meteor the size of Manhattan. The force of the impact would have caused a global holocaust of fire, steam, and dust—creating the equivalent of a nuclear winter. They estimated that up to 90% of the world's biomass had burned. As plants died from lack of sunlight, small animals would have suffocated to death from thick clouds of dust and ash. The dinosaurs and other survivors of the initial effects of the blast perished from the ensuing cold and shortage of food.

Initially scientists rejected the theory because it reintroduced the idea of catastrophism. Most geologists believed that catastrophic cosmic bombardment of Earth had ended long before the dinosaurs, and that slow processes like erosion were responsible for changes in geology and environment. But Alvarez was vindicated when the tell-tale iridium layer showed up in similar layers throughout the world. Recently geologists searching for oil deposits off the Yucatan Peninsula located an underwater impact crater—exactly the size and shape predicted by Alvarez and dating to the time of the Cretaceous extinction.

Extinctions occurred with some regularity, and interest has grown in the possibility that collisions with cosmic objects may have caused other global catastrophes. New evidence links the Permian die-off to an extraterrestrial impact. But two major impacts around thirty-five million years ago, one in Siberia and the other in Chesapeake Bay, appear to have had no effect at all on global ecology. There is also evidence that the global extinctions may not have been as abrupt as once thought. Fossil evidence reveals that some groups of animals perished almost instantaneously, while others died out gradually over millions of years. For example, the dinosaurs may have been in decline for as long as eight million years before the *coup de grace* of an impact resulted in a foregone conclusion.

Today most scientists believe extinctions are caused by a coincidence of survivable events such as global warming or cooling, cosmic impacts, and volcanic activity that, taken together, spell doom. Whatever the ultimate cause of a mass extinction, the consequences are the same. The world's ecosystem collapses. Natural disasters in one part of the planet initiate a ripple effect, and the die-off cascades across boundaries separating organisms. All groups are affected, from mollusks to mammals, dinosaurs to fish. Habitats are destroyed and niches open up for the opportunistic.

Current evidence suggests that 99% of all species that ever lived are now extinct. In other words, the ten million species of plants, animals, and bacteria on Earth today owe their existence to the evolutionary potential of the 1% that managed to survive.

Recently scientists have estimated that the current rate of extinction is between one hundred and one thousand times greater than before the dawn of modern humans. This rate closely matches that of the five global extinctions, and has suggested to some that we are in the midst of a sixth mass extinction.

If you agree with Shaw that there must be a purpose or direction to evolution, how would the reality of mass extinction fit into your view of where we came from?

All her life Anne B. has been teaching others about the laws of physics. Now retired and in her seventies, she believes that evolution must be the Grand Experiment.

If I think about all the plants and animals on Earth and its physical laws, it seems to me there must be a plan. Sometimes I think it all started as a grand experiment, to see where it might go. So I'd call the organizing force The Great Experimenter.

Some people believe in a plan with every detail prearranged. I don't, because in most experiments you may have an overall object, an aim, and a general plan of how to get there, but you don't know for sure what the end result will be.

Look at basic gas theory, where you have thousands of gas molecules moving around in a closed system with a piston. When you heat the gas and the molecules speed up, not all the molecules will hit the piston. But on average enough of them do and the piston moves.

I think extinctions are bumps along the road. Like the molecules that don't hit the piston, the overall trend, the direction, the intent is the same. But not everyone marches in that direction.

Anne is a Moderate who fully accepts modern science and the natural evolution of humans, life, and the Universe. But she makes room alongside evolution for her abiding sense of the spiritual through personification of an ultimate, detached cause. For lack of a better word she labels it the Great Experimenter.

The Rhythm Of Evolution

Like young Don Quixotes, paleontology's best and brightest have often searched in vain for evidence of Darwin's insensible gradations—those incremental forms leading from one form of animal to the next.

Consider the case of Niles Eldredge, a graduate student at Columbia in the early 1960s who was determined to demonstrate Darwin's concepts of variation, adaptation, and speciation in the fossil record of a single animal. He chose the trilobite, an inhabitant of ancient seas whose fossils are abundant, accurately dated, and well preserved in gray shales and limestone outcrops from Illinois to New York (Figure 5.7).

Eldredge assumed he would find gradual, inevitable change.

Instead, he found no change. For millions of years his trilobites had been at an apparent evolutionary standstill. But as Eldredge persevered, he was drawn to the animal's huge compound eyes, each one an aggregate of multiple lenses arranged in precise patterns. It was in those amazing eyes that he finally chanced upon the steps of evolution. Over time a gradual change in the number and pattern of lenses in the trilobite eye had generated new trilobite species. But the changes Eldredge found had taken place in only five to ten thousand years, a geological blink of the eye.

Eldredge had discovered something far more important than Darwin's insensible gradations. He had revealed evolution's rhythm—long periods of evolutionary gridlock punctuated by relatively rapid change.

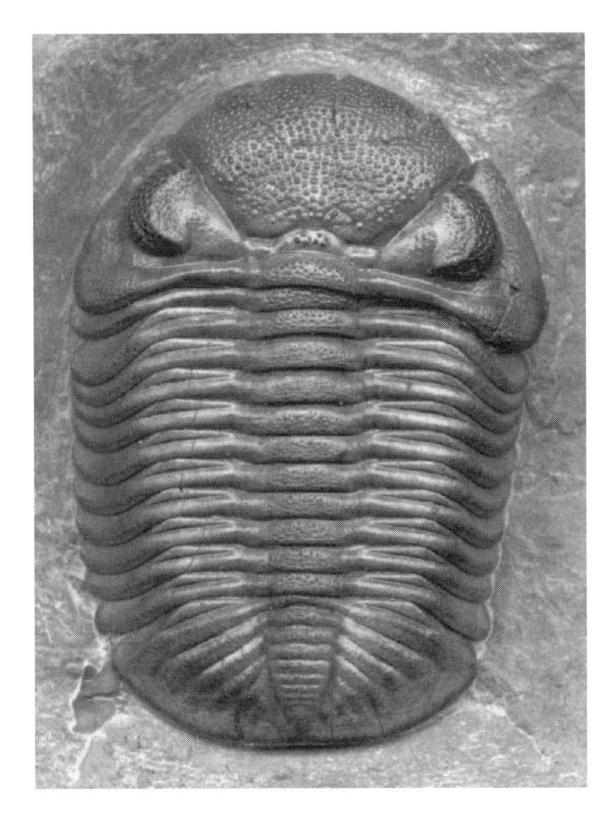

Fig. 5.7 Fossil of a 380 million-year-old trilobite, ancient an-cestor of modern spiders and the equivalent of today's lobsters, shrimp, and crayfish. (Courtesy Samuel M. Gon III)

Stephen Stanley, a paleobiologist at Johns Hopkins, has substantiated the widespread occurrence of this phenomena throughout the fossil record. Stanley and others have shown that stability of form, not change, is the norm. Most change takes place close in time to the original formation of a species and occurs rapidly over thousands, not millions, of years during a very small percentage of a species' lifetime.

Based on these findings, Eldredge and Stephen Jay Gould introduced a new concept in evolutionary theory called punctuated equilibrium. The effect of punctuated equilibrium shows up in the fossil record as long periods of time during which few innovative new features are seen, punctuated by the relatively sudden appearance of new species. Since these bursts of evolution take place quickly around the time a new species is formed, it's unlikely that all intermediates will be preserved as fossils and there may be gaps in the record. Abracadabra? No, just evolutionary sleight of hand!

The Evolutionary Power of Extinction

Now for the good news. Every large-scale extinction has been followed by a burst of evolution. Opportunistic survivors quickly fill niches vacated by extinct species. The rates of change and diversity increase as the number of new species rises exponentially. These evolutionary bursts are the flip side of extinctions. They generate large numbers of new animals and build new ecosystems in short time frames.

The full impact of the evolutionary power of extinction can be seen in the dilemma of the dinosaurs. *Eoraptor*, the earliest known dinosaur, roamed the Earth over two-hundred-twenty-five million years ago when dinosaurs were still small, primitive, and few in number. But within ten million years dinosaurs were flourishing thanks to the late Triassic extinction which hastened the demise of their fiercest competitors, the giant reptiles. Dinosaurs rapidly diversified into niches vacated by the reptiles, and soon became the dominant land vertebrates. After a successful reign of a hundred and fifty million years, the mighty dinosaurs vanished. We could say that dinosaurs were both beneficiaries and victims of extinction.

The evolution of humans can also be credited to extinction. The Cretaceous extinction not only wiped out the dinosaurs, but ushered in an enormously successful diversification of mammals. Within ten to fifteen million years nearly all orders of modern mammals appear in the fossil record, including the earliest ancestors of our own group, the primates.

If species spend most of their lifetime at an evolutionary standstill, what triggers these rapid bursts of change following extinctions? Recent experiments have demonstrated that some genes are hypermutable and show an increased mutation rate in response to environmental stress. We have also seen how mutation in a single gene can reverberate through many others nearby. In some organisms these surplus mutations appear to be stockpiled. Their protein products remain inert or neutral to the animal's survival until a change in cellular environment renders the protein active. The result could be a lethal alteration or natural immunity to new disease.

HIV, the virus that causes AIDS, can live and multiply only by infecting and destroying the T4 killer cells of our immune system. As the virus overwhelms the T4 cells, a cascade of illnesses follow, usually leading to death. In approximately 5% of all HIV-positive cases there are no symptoms of AIDS-related illness, although the victims have harbored the virus for ten years or longer. Studies of these fortunate few have shown no common basis for their long-term survival. But in rare instances, survival can be directly traced to a single genetic mutation coding for a protein that sits on the outside of the T4 cell membrane. HIV must attach to the T4 membrane at that protein site in order to infect the cell. A slight modification in protein structure prevents the virus from docking on the cell and penetrating it. This extraordinarily protective mutation may have existed for millennia in the gene pool of a small fraction of the human population. But it remained neutral in terms of survival until challenged by the new stress of HIV.

It's the Ecology, Stupid

Darwin focused on the importance of the individual in evolution. But when we survey the fossil history of the coming and going of life on Earth, we detect a coarser grained pattern of trends and influences operating at levels above the individual. Events like mass extinctions and evolutionary bursts point to a different set of dynamics that work on species and higher groups. Steven Stanley has studied the pattern of extinctions and uses the term "species selection" to describe interactions responsible for the variable survival or death of entire groups of animals. These higher level interactions create the patterns of macroevolution seen in the fossil record.

Some species may simply be more susceptible to certain types of environmental insults. The Cretaceous extinction wiped out dinosaurs, but not environmentally sensitive animals like frogs and turtles. A weakness might not show up for millions of years in a highly successful species until the chance confluence of environmental and ecological events precipitate disaster.

It may seem obvious that the relationship between animals and their environment will change during global disasters. But interactions among animal species also change dramatically when confronted by a rapidly changing environment. Paleontologist Elizabeth Vrba's studies of the modern African antelope and its fossil ancestors has led her to propose new answers to species extinction and survival. When habitats are destroyed, Vrba says, species must either lead (adapt to the new environment and new mix of species), follow (migrate elsewhere in search of the old kind of habitat), or get out of the way (die). The immediate path a species takes is based on inherited behavior and physiology. In a new environment with a different set of predators and food sources, traits once neutral in terms of natural selection could now become life saving. For example, the untapped ability to digest very coarse grass could enable a minor antelope species to survive and even dominate a normally lush environment that turns arid. Other species lacking this ability might search in vain for greener pastures, only to slowly dwindle in numbers and die out.

Five New Ways to Think About Evolution

Since Thomas Kuhn first introduced the concept of *paradigm* in his 1970 classic, *The Structure of Scientific Revolutions*, we have heard its mantra throughout business, science, and the arts. We throw out the old paradigm and bring in the new for everything from corporate management to suburban architecture. But most people remain unaware of the hold paradigms have on our lives.

They are the implicit beliefs and perceptions that structure our reality and influence the actions we take. In everyday terms, our paradigms determine whether we drive to work or take the bus, whether we vote Republican or Democrat, or whether we vote at all.

Paradigms are unusually impervious to facts. Most attempts to persuade people to abandon beliefs grounded in their personal paradigms will fail, even when those ideas and conclusions have little basis in fact. We abandon our paradigms only by dire necessity. This tenacity is no less evident in the world of science where all endeavors must rely on the prevailing paradigm. In evolutionary science, that paradigm has long been Darwin's theory of evolution—gradual, incremental, continuous, and focused on individual survival. But evolution isn't what it used to be. New research and new insights are fundamentally changing that paradigm. Our picture of evolution must change along with them in five key areas:

1. ***Evolution is episodic, not steady and continuous.***
   For most species, change takes place rapidly, followed by an interval when evolution virtually stands still.
2. ***Evolution progresses through experimentation along many paths.***
   The result is an evolutionary bush that has been pruned by natural selection and chance catastrophe throughout Earth's history.
3. ***Evolution doesn't always rely on the slow accumulation of many small changes. Innovative change leading to unique animals can occur with relative speed.***
   Evidence from genetics and molecular biology reveals that major genetic innovations are fostered by the actions and arrangements of the genes themselves.
4. ***Evolutionary success means flexibility, not complexity.***
   Evolution is the constant generation and testing of designs for life on Earth. Ancient single-celled bacteria

are structurally simple, but their ability to adapt and survive easily in a wide range of environments is a mark of their evolutionary superiority. Genetic variability is the key to survival and reproduction.

5. ***Evolutionary pressures act on entire species and groups, as well as individuals.***
   The result is a selective sorting that produces the larger patterns of survival and extinction we see in the fossil record.

Despite Huxley's famous comment that the slaying of a beautiful hypothesis is the great tragedy of science, one ugly fact won't kill a prevailing theory. But unanswered questions and anomalies that don't fit it will spawn new avenues of research. Alternative theories supported by evidence will form the basis for a new paradigm—a revolutionary way of thinking about the whole subject. In other words, if things don't fit, it's time to change the analogy. This is the true nature of scientific progress.

Kuhn suggested that new and revolutionary theories in science are heralded by "numerous partial solutions that concerted attention to the problem has made available." If this is true, then evolutionary science is flourishing today as never before.

In the twenty-five years since *punctuated equilibrium* was added to the evolutionists' vocabulary, fruitful collaborations among paleontologists, geneticists, ecologists, and others have recast the roles of change, adaptation, and natural selection in light of broader events such as mass extinctions and evolutionary bursts. Vrba, Stanley, Eldredge, Wills, and Gould, among many, have embraced anomalies such as fossil gaps and living fossils, and extended Darwin's theory with new insights. Their work questions what we always thought we knew about evolution and puts new flesh on old bones. Proof of the importance of their effort is the intense discussion it provokes with traditional Darwinian gradualists whom Eldredge, in his book *Reinventing Darwin*, terms the Ultra-Darwinians.

But make no mistake about it. Science has not abandoned the theory of evolution.

The cornerstone of Darwin's theory remains intact. Science isn't questioning that evolution occurs, that species change in time solely through natural laws, or that all life is linked in a pattern of common descent. Instead the scientific debate revolves around the detailed mechanisms and dynamics of evolution. As Eldredge wrote in his book *Time Frames*: "That evolution occurs no biologist worthy of the name doubts. But many biologists these days do openly wonder how complete and accurate our grasp of the mechanics of the evolutionary process really is."

Unfortunately this healthy scientific scrutiny of Darwin's theory has been misread by religious fundamentalists as evidence that Darwin is dead, or at least wounded. During the 1980s, advocates of special creation and intelligent design challenged evolutionists with their own brand of pseudoscience known as "creation science." This attempt to cloak religious belief in the guise of science has had far reaching sociological implications. Though the main focus of this movement has been to advocate the teaching of biblical creation as an alternative to evolution, its arguments have captured undue media attention because they strike at the ABCs of the evolution controversy.

REFERENCES

Dawkins, Richard. 1986. *The Blind Watchmaker*. New York: W. W. Norton.

Eldredge, Niles. 1989. *Time Frames*. New Jersey: Princeton University Press.

__________. 1995. *Reinventing Darwin: The Great Debate at the High Table of Evolutionary Theory*. New York: John Wiley.

Huxley, Thomas Henry. 1896. "Biogenesis and Abiogenesis." *Discourses Biological and Geological*. New York: D. Appleton.

__________. 1901. "Letter to Darwin (1859)." *Life and Letters of Thomas Henry Huxley*. Leonard Huxley (ed.). New York: D. Appleton.

Kuhn, Thomas S. 1970. *The Structure of Scientific Revolutions*. Chicago: University of Chicago Press.

Peck, M. Scott. 1978. *The Road Less Traveled*. New York: Simon and Schuster.

Shaw, Bernard. 1947. *Back to Methuselah: A Metabiological Pentateuch*. London: Oxford University Press.

WEBSITES

University of California Museum of Paleontology (http://www.ucmp.berkeley.edu) Fossils and more on this terrific website sponsored by the largest fossil collection in any university museum in the world. Great graphics and links.

A Guide to the Orders of Trilobites (http://www.aloha.net/~smgon/ordersoftrilobites.htm) Only the worst mass extinction in history could stop the 300 million year run of the trilobites—the most diverse animals in the fossil record. Award-winning site by Sam Gon III, Hawaiian biologist and trilobite enthusiast, will get you hooked on fossils and the nuts and bolts of evolution.

SUGGESTED READING

Climbing Mount Improbable, Richard Dawkins, W. W. Norton, New York, 1996. Engagingly written by a leading Ultra Darwinist who argues that gradual change has the power to build complexity in nature. Wonderful illustrations.

Reinventing Darwin, Niles Eldredge, John Wiley & Sons. New York, 1995. Lively introduction to the leading edge of evolutionary theory today. Well-written, non-academic exploration of all sides of the scientific debate by an insider. For everyone interested in evolution and the story of life on Earth.

Chapter 6

Once Upon a Time

Life is what happens to you while you're busy making other plans.
John Lennon

Several prominent biologists and chemists were meeting at a large university to discuss the significance of a new discovery involving the molecular structure of genes. They were interrupted by a secretary with a request from the local philosophical society for a speaker to participate in an upcoming seminar on the nature of life. The scientists looked at one another rather perplexed, then politely declined the invitation. None of them thought they had anything to say about life.

It seems that the more we learn about life, the less eager we are to define it.

Science has taken an end run around the question by dicing it into manageable, testable pieces. We have a list of what life *requires* —carbon, oxygen, water, and energy. We also have a working definition based on what life *does*—replicates, metabolizes, mutates, and undergoes Darwinian evolution. But life is more than carbon molecules that crawl.

What is life? It depends on who you ask and what your definition of the word *is* is. Astronomers scan the heavens and confirm that the basic ingredients for life exist throughout the stars. Many scientists believe there is a universal chemistry of life. But so far the only example we have of it is life on Earth.

Biochemists are able to demonstrate how the first molecules of DNA might have assembled into a blueprint for life. They have decoded the genetic alphabet, sequenced all three billion letters in the human genome, and are busy translating its story. Paleontologists and geologists examine rocks and find the oldest fossil and chemical traces of life on Earth. Microbiologists search for life and find it in all the wrong places. From Yellowstone's hot sulfurous springs to fresh water lakes deep beneath the Antarctic ice, they capture microbes that eat iron and are able to thrive without oxygen. These organisms may have been our earliest ancestors. Exobiologists study these extreme life forms in order to recognize similar life on other planets. But the fact is that we still don't know how life began on Earth.

What does it mean to be alive? Most formal definitions of life leave much to be desired. If we say life is something that replicates, changes, and undergoes natural selection, we have to include fire and mineral crystals, but leave out Aunt Betsy who can't have children.

Viruses are submicroscopic cellular intruders. Stripped of all but the bare essentials of a handful of genes and a protein coat, viruses hijack their victim's metabolism in order to survive and reproduce. They can't do any of that on their own. Though scientists debate whether viruses evolved before cells or descended from them, they agree that viruses are alive.

But consider *prions*, misshapen bits of protein that cause normal proteins to warp and form insoluble clumps in the brain. Prions are responsible for fatal diseases like mad cow disease and Creutzfeldt-Jakob disease in humans. An article in the London *Daily Telegraph* proclaiming "Humans may be secret carriers of mad cow bug" alerts us to the possibility that prions from infected cows may have slipped into the food chain during the 1980s mad cow epidemic in Britain. Once considered harmless to humans, these mad cow prions have been able to mutate and jump the species barrier. They are now blamed for a variant of Creutzfeldt-Jakob in humans. Prions may lie in wait for years, undetected in otherwise healthy animals like the silent reservoir of a deadly epidemic.

Where and how did life begin? Spontaneous generation is probably the oldest explanation for a natural origin of life. Aristotle wrote that insects and worms could emerge spontaneously from non-living matter such as dirt, dry leaves, or the remains of dead animals. Seventeenth century British anatomist William Harvey, celebrated for his revolutionary studies of the human circulatory system, considered spontaneous generation to be a common means of animal reproduction alongside sexual and asexual methods. But he insisted that in each instance the ultimate source of life is a "vegetative force," a spark of life planted by God into non-living matter.

In 1864 Louis Pasteur first successfully challenged the ancient and universal belief in spontaneous generation. In his classic experiments containers of nutrient broth were either left open or tightly sealed in the dusty laboratory air. The open flasks soon became contaminated with bacteria and mold, while the sealed containers remained sterile. These results led him to conclude that "there is no known circumstance today in which one can assert that microscopic beings have originated without germs."

But beyond the simple question of life's biological origins lies a host of philosophical problems. Is life a miracle or a rare but natural occurrence? When certain conditions are met, the known laws of chemistry and physics warrant that life is not only possible, but perhaps a routine event. Could life simply be the natural consequence of an orderly Universe and its penchant for self-organization?

Arthur Peacocke, biochemist and Anglican priest, agrees that science has established that life is inevitable and will occur under certain environmental conditions. A "God of the gaps" isn't needed to fill in every missing link in evolution. But Peacocke's theological view is that science has established only that life will naturally emerge somewhere in the Universe. The details have been left open to interpretation and there is still plenty of room for God.

Kathy L. is a seventy-year-old retired statistician. Her analytical mind grasps the nature of chance, but her psychology of belief yields to an unseen hand in the calculation of life.

Bible stories are allegorical. They're meant to teach us how to live. I agree with Thomas Jefferson. He was a deist and said that the world was created by a divine being who does not take a day to day personal interest in our lives.

We evolved. It's as simple as that. God made it happen. The Universe is pulsing with life. We're not the only planet with life. Why would God create all the other planets and worlds and leave them dead?

Kathy believes that life developed naturally, but God was the Ultimate Mover and Shaker. This view is typical of many Moderates. Back here on terra firma we are searching for the holy grail of life's origins—that first replicating molecule. Some scientists think they have found it. What were the conditions on Earth when one moment there was a molecule, the next there was life?

It Was a Bright and Stormy Night

Lightning raced through the dusty orange sky and a full moon covered the horizon as far as the eye could see, shedding light on the bubbling inferno below.

Martian landscape? Alien world? No, it's planet Earth four and a half billion years ago. Startling new evidence is altering our picture of the Earth's early days. Cosmologists tell us that the Universe was born in a Big Bang roughly fifteen billion years ago. Compared to the Universe, our Sun is a mere teenager born 4.6 billion years ago out of the remnants of long dead stars. Gravitational forces in a cloud of interstellar dust compressed this star stuff into a disk so dense that it exploded into an atomic fire that still warms our planet. The young Sun's energy blew stellar material farther out where it collided and coalesced into giant gaseous globes and rocky spheres. These solar step children raced and collided among themselves, eventually settling into the planetary system we recognize today.

By four and a half billion years ago, geophysical forces had created an Earth of striking contrasts:

- An object three times the mass of Mars had slammed into Earth, setting it spinning on its axis so fast that a day was only two hours long.
- An enormous Moon, formed from rock launched into orbit by that collision, hung only fourteen thousand miles above our planet.
- Constant bombardment by comets, asteroids, and meteorites fueled the molten planet.

The rain of celestial trash continued for a billion years, vaporizing rock, boiling oceans, and contributing heavy elements, light organics, and water to our developing planet. With a Moon too big, a Sun too faint, no bacteria, plants, or animals, and no oxygen to breathe, primitive Earth would have been as alien to us as Mars.

If we could travel back in time and hover over our planet four billion years ago, we would see:

- fourteen-hour days as the Moon's gravity slows Earth's spin
- a receding Moon still appearing four times larger than today
- an ocean with mile-high tides, and sunshine 30% dimmer than now
- toxic clouds of carbon-dioxide and hydrogen sulfide lending an eerie orange glow to the sky
- the first traces of free oxygen in the atmosphere—and perhaps the beginnings of life.

Recently the Hubble telescope captured dramatic images of an erupting volcano on Io, Jupiter's closest moon. Io is about the same size as our Moon and is the most volcanically active body in our solar system. Its powerful volcanoes spew two-hundred-fifty-mile-high plumes of dust and gas that reach temperatures of twenty-seven-hundred degrees Fahrenheit—creating extreme environments similar to those on Earth four billion years ago when life first began.

Throughout millions of years our planet was covered by a single ocean dotted with small volcanic islands. No large landmasses lingered. Earth's volcanic conveyor belt devoured rock, melting and remaking crust and obliterating any clues to early life.

Around 3.9 billion years ago a dramatic increase in cosmic impacts resurfaced the lunar landscape and affected the entire inner solar system. Giant asteroids pummeled and sterilized our planet. Life on Earth may have been snuffed out many times. But tantalizing traces of the presence of cells in rocks from Greenland that date to over 3.8 billion years ago suggest that Earth may have harbored life in the midst of this cosmic rain of terror.

By three and a half billion years ago the bombardment had slowed, and we see what may be the oldest evidence of life on Earth—thread-like fossils of single-celled organisms resembling bacteria. Their diversity and complexity suggest that life had been evolving for some time.

Today we live in a world of great beauty and stunning diversity. But simple microscopic life filled the oceans with green slime and dominated life on Earth for most of our planet's history. Mutations in these bacteria-like organisms produced offspring that gave off oxygen as a by-product of metabolism. Other bacteria were able to survive by using this oxygen to break down food. Primitive single-celled plants gave off oxygen into the atmosphere, and eventually an ozone layer formed that protected life from harmful UV radiation.

Around two billion years ago:

- A build-up of large amounts of atmospheric oxygen launched the evolution of more complex cells.
- Sex appeared (among single cells, that is) coinciding with a surge in multicellular life. With the mixing of genes from two parent cells came new combinations, unique adaptations, and evolutionary advantages.

By one billion years ago:

- Cells teamed up and specialized to became the first multi-celled organisms.
- Some cells enlarged and packaged their genetic material into a nucleus. They engulfed oxygen-using bacteria whose remnants are with us today as mitochondria, the cell's energy factories.
- These more complex modern cells spawned a surge in evolution that eventually gave rise to the first multicellular animals.

The Origin(s) of Life

Some scientific theories read like science fiction.

In 1924 the Russian chemist Alexander Oparin suggested that life emerged from a rich primordial soup of minerals and organic molecules. Organic gels produced by this mixture had enough energy to carry out the simple chemical reactions that boosted them into the realm of the living. Oparin's slime "could not metabolize in the full sense of the word, [but] it must certainly have had the ability to nourish itself, to absorb and assimilate substances from its environment."

Many scientists latched on to Oparin's vision, theorizing that life could have begun in a chemically reactive sludge, warmed by the sun and mixed by the winds of a primitive planet. But American chemists Stanley Miller and Harold Urey were the first to attempt to recreate the origin of life in a test tube.

In 1953 Miller filled a large round flask with boiling water and three gases (methane, ammonia, and hydrogen) to simulate Earth's primeval atmosphere. He sealed the flask and zapped it with an electric charge equivalent to a bolt of lightning. In a scene reminiscent of the 1950s movie classic "The Blob," Miller returned to his lab a week later and discovered that a red tar-like substance had formed in the flask. The residue proved to be a rich mixture of two simple, but important, amino acids found in the proteins of all living organisms. When Miller prolonged his experiment or varied the conditions, more complex amino acids formed. Miller's daring experiment proved two things. First, under certain conditions, the complex building blocks of life can easily assemble from non-living matter. Second, it is possible to test a scientific hypothesis about the chemical origins of life.

Unfortunately, Miller got the recipe wrong. We now know that the early atmosphere of Earth was deadly. Rich in carbon dioxide and nitrogen, it easily destroyed most organic molecules as soon as they formed. The concentration of the precursors of life available at any one time was probably very low. The sea of life was more like a cup of bouillon than chicken soup.

There is also growing evidence that significant amounts of the chemical precursors of life rode to Earth on asteroids, comets, and meteorites. For example, when comet Hale-Bopp recently sailed through our solar system, it melted and shed molecules of complex organic compounds along its path. Every day hundreds of tons of dust from dark interstellar clouds fall to Earth, furnishing many different types of organic chemicals. Putting it another way: "I am bidden to surrender myself to the Lord of the Worlds. He it is who created you of the dust." (The Koran, Sura 40)

This interstellar delivery system may have supplied Earth with many of the raw materials of life. Recent discovery of a simple sugar compound in space raised the odds that more complex precursors of life are abundant throughout the Universe.

Jean G. is a thirty-something physical fitness instructor with a large following of students. Her vivacious personality and carefree attitude mask more serious spiritual convictions.

We were created by God through an evolutionary process that has taken billions of years to complete. The biggest unknown about life is the state of life prior to the beginning of the Universe. Life has always existed because God has always existed.

For folks like Jean, the origin of life is a "no brainer." But for others, the devil is in the details.

At some point during Earth's hellish gestation, selective pressures overcame accidental chemistry. That first living molecule may have been produced in a single-step evolutionary event. What was it? How did it develop? Today researchers are tackling this hot topic from several angles.

- Gerald Joyce, a biochemist at the Scripps Institute in La Jolla, believes "You can't put forward a firm scientific definition of life. It's a term that really has only popular meaning." He should know. Joyce has come the closest to creating molecules that imitate life. Joyce and others believe that when life first began, it was an RNA world. The first self-replicating molecule wasn't DNA, today's genetic star, but RNA—a large molecule similar to DNA that performs some of the most ancient and important processes in the cell. For example, RNA reads and copies the instructions coded in our genes and translates them into specific proteins that do the work of the cell. Under the right conditions RNA can copy itself. Gerald Joyce's RNA molecules not only replicate, but also change and undergo a type of natural selection. In other words, they evolve.

The origin of life on a molecular level also means the origin of coded information. But RNA and DNA are large, unwieldy molecules, and very difficult to make. How did these highly ordered molecules of life form in the first place?

- Graham Cairns-Smith believes he has the answer. In his book *Genetic Takeover and the Mineral Origins of Life*, Cairns-Smith suggests that mineral clays abundant on our primitive planet acted like a tinkerer's workbench to sequester scarce carbon compounds and align them to promote reactions that created the first replicating molecules. Recent experimental evidence suggests that mineral clays can stimulate the synthesis of RNA and DNA, lending support to Cairns-Smith's interesting theory.

- In contrast, biochemist David Deamer still envisions the origin of life as a fragile, tenuous event occurring in the womb of warm tidal pools where fatty substances curled into cell-like bubbles like froth on the beach. Deamer believes that these microscopic structures concentrated scarce organic materials and encouraged the fragile bonding needed to make RNA and DNA. Cell membranes might have assembled the same way.
- Biophysicist Stuart Kauffmann also pictures the origin of life as the inevitable consequence of nature's unlimited potential for self-organization. Snowflakes, patterned crystal growth, and Deamer's fatty bubbles would be prime examples of Kauffmann's concept of the hidden power of self-ordering systems. He suggests that natural selection simply added more layers of complexity and efficiency to this inherent biological harmony.

There is growing evidence that life on Earth developed very fast under foul conditions. The oldest known cells were already using complex chemical processes to convert carbon-dioxide into energy-rich sugars. This suggests that the very first life forms may have employed a simpler but unique chemistry to survive in a chemically chaotic world.

- Chemist Gunter Wachterhauser agrees, claiming that life arose not in a warm pond, but in the chemical chaos of undersea volcanic pressure valves. Like a mad chemist's pressure cooker, undersea hydrothermal vents mix carbon and hydrogen-rich gases with metallic sulfides abundant on the ocean floor. The result is a brew of compounds that catalyze the manufacture of

> proteins and cellular building blocks. Traces of this ancient chemistry may survive today in Archaea—ancient forms of cellular life that thrive in boiling hydrothermal vents and live deep inside solid rock by using chemical energy instead of sunlight. Archaea are the oldest and most primitive forms of life on Earth and may be our direct ancestors. They resemble bacteria, but are as genetically different from bacteria as we are. Archaea and hydrothermal vents might lead us to a universal common ancestor of life—that first cell-like critter.

But not everyone agrees that life began on Earth. Life appeared at the first instant it could, but already exhibited signs of sophisticated chemistry and complex organization. Moreover, all life on Earth incorporates the same unique set of twenty amino acids in its proteins and relies on the same cellular and genetic information systems. This suggests to some very smart folks that life as we know it didn't begin here but on another world under different physical and time constraints. If so, did life arrive on Earth by accident or intent?

In the early 1900s, Swedish physicist and Nobel prize-winning chemist Svante Arrhenius coined the term *panspermia* to express his belief that microscopic seeds of life exist throughout space and constantly rain down on Earth. More recently, Nobelist Francis Crick (of Watson and Crick double-helix fame) has taken up the banner of panspermia, suggesting that living microbes rode to Earth in a space vehicle sent from a civilization in another galaxy, long ago and far away.

It's not as far fetched as it sounds. We know that the chemicals of life abound in the cosmos and that life is tenacious. Studies have shown that microbes survive long periods in space. Bacteria excel at forming non-vegetative spores that survive under toxic conditions and flourish once again when conditions return to normal. If replicating molecules or rudimentary cells had evolved on other worlds, they might have survived a trip to Earth on pieces of asteroids or bits of dust. If these interstellar missionaries had been intentionally directed our way, as Crick suggests, their survival would explain the selective chemical hand that life on Earth has been dealt.

The Dark Star

Ancient tribal lore of the Dogon people of Africa describes an invisible star that was the source of all life on Earth. Only in recent years have astronomers detected a collapsed or "dark star" located in the area of our galaxy where sacred Dogon drawings place it. Is this evidence for the extraterrestrial origin of life or just coincidence? Is anyone else out there?

A recent ABC News poll reported that 69% of Americans believe there is intelligent life on other planets. Thankfully, only 1% of the poll's respondents said they had been contacted by the aliens. Arthur C. Clarke, author of *2001: A Space Odyssey*, summarized the dilemma saying, "Two possibilities exist: either we are alone in the Universe or we are not. Both are equally terrifying."

Early Greek philosophers known as Atomists believed that there were many worlds and many universes that could be inhabited. But Aristotle claimed that the world we know is the only one that exists. The 1st century B.C. Roman philosopher Lucretius returned to the Atomist tradition when he wrote, "Space contains such a huge supply of atoms that all eternity would not be enough time to count them. So we must realize that there are other worlds in other parts of the Universe, with races of different men and different animals." Only four hundred years ago the Italian astronomer Giordano Bruno was burned at the stake by the Catholic Church for suggesting the same thing.

The chance of finding life on another planet may be pretty good. We know that the Universe is fifteen billion light years across and contains at least 100 billion galaxies, each consisting of billions of stars. Surely intelligent life lurks on another Earth-like planet revolving around a star like our Sun among the 400 billion stars of the Milky Way Galaxy. Yet the search for extraterrestrial life has long been considered the province of kooks at the fringes of science.

The late Carl Sagan championed the Search for Extraterrestrial Intelligence or SETI—a rigorous scientific program that monitors interstellar radio frequencies for signs of alien communication. Since 1989 six suspect signals have been detected, but none have been repeated. In the latest version of "E.T. phone home," SETI will soon optically scan the sky for laser pulses as brief as one billionth of a second, signs that up-to-date aliens may be calling Earth.

In 1938 Orson Welles' realistic radio broadcast of H. G. Well's science fiction classic *The War of the Worlds* sent Americans scrambling for cover from invading Martians and rekindled our fascination with life on other planets. We were rewarded in August 1996 when NASA scientists announced they had found evidence for ancient life in a 3.6 billion-year-old meteorite from Mars.

But the exhilaration was short-lived. Within two years NASA's evidence had crumpled. Traces of amino acids similar to those found in the Martian meteorite also showed up in the surrounding Antarctic ice. Magnetite crystals seen in the rock and thought to be the chemical calling card of ancient life were found to have non-organic origins. Minute cell-like structures in the rock were shown to be anomalies. Most damning of all was the fact that other non-Martian rocks, including those brought back from the Moon where no life exists, also showed some of the same "evidence" for life.

Will we recognize alien life when we see it? Jan B., a forty-year-old housewife from the Mid-west, has her doubts.

Life on other planets seems far fetched to me. But I don't see why not. I think we're projecting our own ideas of "being" onto what life might be like on other planets. It could be very different like George Lucas imagines it. Aliens would need a Rosetta Stone to understand us.

Fallout from the Mars meteorite fiasco prompted NASA to assemble a team of experts to create their own "Rosetta Stone"—a list of chemical compounds that might indicate the presence of alien life.

Could life have survived a ride to Earth on an asteroid or loose chunk of planet? NASA scientists have revived the old theory of panspermia and redefined it as interplanetary transfer of life. Researchers are testing whether hardy microbes that thrive in the nastiest environments on Earth could survive an interstellar trip. Their studies suggest that the chances of cellular life surviving the radiation, heat, and impact forces associated with interstellar travel are slim to none. There is a much higher probability that viable transfers of cellular life could have taken place between planets within our own solar system or from nearby solar systems of similar age.

Recent discoveries have brought science fiction closer to reality. Since 1995 astronomers have confirmed the existence of over seventy planets orbiting fifty sunlike stars in our Galaxy. At least six of these stars have multi-planet systems. Some of these new worlds are close enough to see. For example, the sunlike star Epsilon Eridani that lies just west of the constellation of Orion is the nearest star known to be orbited by at least one and possibly several planets. This star system is only 10.5 light-years (sixty-one trillion miles) away and is visible to the naked eye.

In August 2001 astronomers announced the discovery of the first distant planetary system in which the planets have near-circular orbits like those in our own solar system. Two gaseous, Jupiter-like planets circle the faint star 47 Ursae Majoris in the Big Dipper. Discovery of a planetary system bearing some resemblance to our own increases the odds that Earth-like planets capable of supporting life may be found.

It's difficult to predict which of these new planets might harbor life because no one knows what the environment was like on Earth when life first secured a permanent foothold. Our planet is located in the habitable zone of the Sun where liquid water is abundant and extremes in temperature haven't impeded the development of complex life. But four billion years ago when life first arose on Earth, the Sun was only two-thirds as luminous as today—guaranteeing average temperatures on Earth well below freezing. Perhaps the greenhouse effect was stronger in the past, or heavy bombardment and volcanic activity kept the Earth warm and cozy for the first living cells. For these reasons scientists have expanded the "habitable zone" in their search for extraterrestrial life to include more radical temperatures and pressures that might foster extreme forms of life.

Liquid water has recently been detected beneath the icy surface of three moons of Jupiter—Europa, Ganymede, and Callisto. Europa's ocean lies only twenty miles below the ice on a rocky floor that may contain hydrothermal vents. This stunning find makes Europa one of two prime candidates for extraterrestrial life in our solar system. The other is Mars.

Images from the Viking Lander survey of Mars in the 1970s, combined with recent measurements taken by the orbiting spacecraft Mars Global Surveyor, have revealed the shoreline of a vast ocean that covered the surface of the planet two billion years ago. Exciting new evidence points to the possibility that substantial amounts of water as subsurface ice may still exist on Mars today. The hope of finding ancient Martian microbes, if not little green men, is alive and well.

In 2001 researchers synthesized primitive cell-like structures by blasting a frozen mix of water and cosmic gases (methanol, ammonia, and carbon monoxide) with enough UV radiation to equal that generated by young stars. The resulting slushy mix of organic compounds formed membranous bubbles in water similar in size to living cells. More exciting was the discovery that these chambers are able to transform UV radiation into wavelengths of visible light needed for life. These proto-cells might have been abundant in the Universe, falling like seeds of life on alien planets. But what is the possibility that *intelligent* life has evolved elsewhere?

Many people, including some scientists, say that the development of complex life on Earth was a singular event. Their evidence ranges from brilliant physics to blind faith. In their book, *Rare*

Earth, astrobiologists Peter Ward and Donald Brownlee argue that simple microbial life may be even more abundant in the Universe than now imagined. But complex life requires precise conditions to take hold and evolve. It may exist in no place except here on Earth.

> Al W. is a rocket scientist, literally. When this middle-aged engineer isn't designing liquid propulsion systems, he's inventing futuristic machines and cozying up to some bedtime reading on quantum physics. Al trusts facts and figures, but his logic takes an interesting twist when it comes to life on other planets.
>
> *I don't believe that complex life exists on other planets. We know that the elements that make up our bodies are manufactured in the gravitational collapse, explosion, and death of stars. So the stuff of life exists throughout the Universe. We also know that human evolution took billions of years. If life is out there on other planets, it probably isn't at the same stage of evolution that we are.*
>
> *As a Christian I believe that Christ's death is sufficient for a "people." If human life exists on other planets at different times, then multiple deaths of Christ would be required for the salvation of all the people in the Universe. I can't accept this multiple death theory, so for me the assumption that life has existed elsewhere in the Universe is false.*
>
> *The Universe is therefore for us, the people of the Earth. Christ died only for the people of Earth because we are the only people in the Universe. We are alone.*

If you followed Al's reasoning you know he's done some heavy thinking on the subject. But the bottom line is that despite his trust in science, Al is even more strongly committed to the literal truth of the Christian concept of salvation.

Traditional religions have an Earth-centered point of view. For many believers like Al, their reference to Christ as the "Light of the World" means this earthly world we know, not the Universe.

How would folks who already have a hard time accepting different human races react to a completely different form of intelligent life? Could we presume that aliens would have an ethical code or value life as we do? As biologist Lewis Thomas describes the scenario in his book *The Lives of a Cell*:

> *The main question will be the opener: "Hello, are you there?" If the reply should turn out to be "Yes, hello," we might want to stop there and think about that, for quite a long time.*

Talking Bones and Molecular Maps

"If it looks like a duck and walks like a duck" has been the guiding philosophy of evolutionists over the centuries—and for good reason. Finding evolution's trail is all about pattern and similarity. Nothing offers more tangible evidence of the history of life on Earth than fossils.

Fossils are traces of earlier life forms that have been preserved in sediments, captured in amber, frozen in ice, or mummified in dry air. Fossilized teeth, jaws, limbs, and shells establish similarities in structure and function that suggest hereditary relationship. When placed in chronological sequences, they reveal patterns of emergence, development, and disappearance of individual traits and entire groups of organisms. Fossils can provide insight into how animals survived in various environments. For example, teeth tell us about diet (meat eater or vegetarian), limbs and hips tell us about size and locomotion (land walkers or tree dwellers), and size and shape of the skull suggest cognitive abilities.

Most fossils come from organisms with hard parts like bones or shells that fossilize if they are buried in sediment before they deteriorate. Minerals in the mud replace calcium in bones, creating rocky jewels in a wide range of colors. Soft-bodied organisms quickly deteriorate and rarely fossilize. But under the right conditions in fine-grained sediment, details of soft body parts will fossilize. When soft and hard parts fossilize together, the result can be a strikingly realistic portrait of life, like that of the fossil of the oldest bird Archaeopteryx. Sometimes whole animals are preserved, such as insects caught in amber resin or mammoths frozen in ice. Trace fossils can form from undisturbed tracks of animals or the burrowing of worms into soft sediment. The amazing 3.7 million-year-old footprints of early human ancestors found by Mary Leakey in Laetoli, Tanzania are examples of trace fossils.

Fossilization isn't a sure thing. Timely burial, a climate that inhibits decay, protection from scavengers, and a lack of geological disturbance increase the odds. Half of all plants and animals in any environment will not normally fossilize. Amoeba don't fossilize, but some bacteria do. The oldest known fossils are stromatolites—layered pillow-like rocks built up from the growth and decay of mats of algae-like bacteria. Even when fossils form, it is unlikely they will survive the forces of erosion and shifting ground. Millions of species have lived and died without leaving a single fossil or any trace whatsoever that they existed.

In the absence of visible fossils, life can leave its calling card in the form of biomarkers—unique chemical compounds made only as a by-product of cellular metabolism. NASA's recent claim for evidence of life on Mars was based in part on a biomarker found in the meteorite and thought to be produced only by living processes.

Sophisticated techniques based on the decay of radioactive isotopes of carbon, uranium, and potassium can provide accurate ages for minerals and organic matter up to one and a half billion years of age. Like peeling an onion, evolutionists combine dating techniques with inventories of fossils found in dated rock strata to uncover the layers of life on Earth in past epochs. But a powerful interloper is muddying the fossil trail, bringing with it controversial new evidence that challenges long held theories. It's called DNA.

The O.J. Simpson trial taught Americans the ins and outs of DNA. Now scientists are using DNA sequences to map the genetic past lives of living organisms. Bones and teeth are only the visible evidence of much deeper genetic connections. Like molecular fingerprints, the similarities and differences in specific DNA sequences of our genes are clues to our evolutionary heritage.

For example, recent excitement surrounding the decoding of the human genome has overshadowed previous unraveling of the genes of simpler organisms such as bacteria and yeast. These life forms diverged from a common ancestor billions of years ago, but have retained genes that are similar in structure and function. By comparing their shared genes plus others essential to cell survival, scientists hope to reconstruct the minimum amount of genetic capital required for the first cell. It could be as few as two hundred and fifty genes.

Scientists have also used DNA to build a "molecular clock" that measures the point in deep time when major groups of organisms or single species emerged. Over time genes change because of normal mutation. By assuming that genes mutate at a relatively constant rate, molecular biologists can measure the amount of change in a similar area of the genome among several living organisms to determine how long ago they split off from a common ancestor. The older the group or species, the more its genes will have changed. The molecular clock assumes that the mutation rate has remained relatively constant throughout evolution. Usually the genes selected for the clock are known to be fairly stable and not prone to a lot of breakage. Sounds pretty logical and reliable so far, right? Well, not if you're a paleontologist.

Paleontologists fear that the molecular clock has missed a few beats here and there. They argue that the rate of mutation may have increased during evolutionary bursts, causing the clock to calculate a species to be older than it is. Or the mutation rate may have unknowingly slowed in some groups of animals before resuming a normal tempo. The molecular clock would compute the evolutionary history of these species to be much shorter than it is. In most cases the clock reinforces the fossil evidence, but not always.

For example, there are no confirmed fossils of adult animals older than six-hundred million years ago when organisms considered to be true animals first appeared. But microscopic fossils of dividing animal embryos found in five-hundred-seventy-million-year-old rocks suggest that eggs of even older animals could have survived in more ancient rocks.

Into the midst of this fossil stand-off comes evidence from the molecular clock that traces the origins of animals back as far as 1.2 billion years ago—nearly twice the age supported by fossil evidence.

It is possible that the very first animals were too tiny and fragile to leave fossils. The DNA clock may be all the evidence we ever have. But the paleontologists prefer the hard evidence of fossils.

Brad J. is a forty-five-year-old high school history teacher and avid reader of popular science magazines. His lifelong interest in science has led him to come to his own conclusions about the limits of science.

Intense emotions do a lot of things to the energy in our minds. I am receptive to the possibility of psychic communication by means of an unknown energy. We have a chance to find an answer to psychic experience by conducting experiments, but the question of origins is something we'll never be able to test.

They're doing cloning, but that's not the same thing as creation. They have to have something to start with. I can believe that DNA formed from the right chemicals in the environment, but you've got to have energy come in at some point. That's what I call the Original Force or the Creator. I don't believe humans can do that. We're using what we have. I have very serious doubts that anyone will be able to show me how they created the original.

In our personal lives as well as the hallowed halls of science, what we believe depends on what we are willing to accept as evidence.

Connecting the Dots

And now for the rest of the story.

We left life on Earth one billion years ago in the throes of a sexual revolution that led to the first animals. From this point forward our picture of evolution remains cloudy for nearly a half billion years.

The stars continued to play games with life on Earth. There is evidence that an enormous meteorite hit the Earth around five-hundred-forty million years ago, and may have kicked-off the final journey from the first animals to modern humans. The earliest fossils of most lines of modern animals appeared then during the Big Bang of biology known as the Cambrian. Multicellular life suddenly exploded in breadth and complexity. We have long known that fossils of complex marine life existed at the dawn of the Cambrian. But during the preceding billions of years during the Precambrian, life had remained, for the most part, single-celled and simple. Abracadabra? Not exactly.

In the 1940s Australian geologists uncovered a motherlode of marine fossils at a site called Ediacara that dates to the late Precambrian between five-hundred-fifty and six-hundred million years ago. These bizarre soft-bodied creatures were a preview of coming attractions. Some were bottom dwellers shaped like large doormats, while others resembled spoked wheels and palm fronds. The smaller, more mobile Ediacarans may have been some of the first members of the Animal Kingdom. Others were fleeting experiments that left no further trace in the fossil record.

Recently a candidate for the world's oldest animal fossil was found in the Mexican desert a hundred miles south of Tucson, Arizona. This two inch long jellyfish-like creature lived in the sea six-hundred million years ago.

So far we've been talking about soft, spongy sacks of cells. But with the rise of the of animals, complexity takes on a whole new dimension. Organization, specialization, and communication among different cells produced limbs, eyes, jaws, teeth, specialized organs, sensory response, and organized thought. We were on our way to becoming human.

Fossil evidence clearly shows that around three-hundred-sixty million years ago lobe-finned fish like the Coelacanth developed limbs. These special fish had gills, plus the added advantage of simple lungs. In times of drought they could drag themselves out of water onto dry land, take a few gulps of air, and move to a better pond. Gradually they developed limbs, toes, and the strong hip and shoulder muscles to support them. Primitive four-limbed animals, like *Acanthostega* (Figure 6.1), developed a skill for walking in swampy land but were primarily aquatic. Eventually a sturdy rib cage evolved to lift their internal organs up above the ground as water once did. About three-hundred-thirty million years ago they branched into two groups. One led to amphibians such as frogs, and the other to all land vertebrates, including reptiles, birds, mammals, and humans.

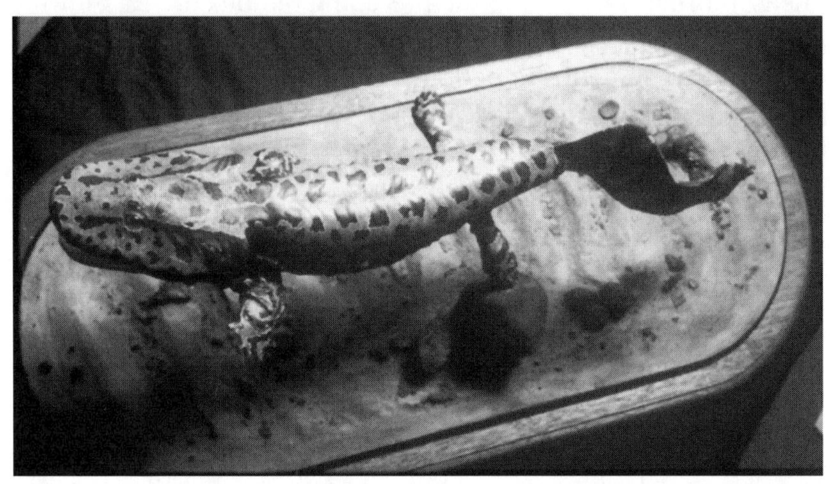

Fig. 6.1 *Acanthostega*, the granddaddy of all land walkers, lived in the swamps of Greenland three-hundred-sixty million years ago. (Copyright Richard Hammond)

Meanwhile in the drama of evolution, not only were the actors wandering, but so was the stage. Around two-hundred-fifty million years ago the Earth's landmass was united into a single supercontinent known as Pangea. Soon tectonic forces split this large crustal plate, creating several smaller continents that wandered from the poles to the tropics. This planet-wide remodeling dramatically altered environments, provoked competition among organisms, and redistributed life in patterns that Darwin tried in vain to explain.

For example, during the age of the dinosaurs around a hundred and sixty million years ago, a large southern continent known as Gondwana began to break up. It eventually formed South America, Africa, southern Asia and Australia. At the time

Gondwana split, all mammals were marsupial like kangaroos. Their young matured in an external pouch. Marsupials predate the rise of placental mammals such as humans, whose offspring develop in a membrane or placenta inside the uterus. Placental mammals first evolved in Asia, then spread around the globe, overcoming nearly all marsupial mammals. But they never reached Australia, or if they did they lost the battle there. Marsupials remained the dominant mammals in Australia, flourishing without any real competition until eight thousand years ago when aborigines from Asia introduced the wild dog they call a dingo.

Today the combination of talking bones and molecular maps is painting a far richer and more detailed picture of evolutionary moments than we could have imagined. For example, paleontologists have long maintained that fossil evidence proves whales evolved from mammals that walked on land. This has been a whale of a tale for a lot of folks who find it hard to envision evolution rewinding itself.

But over the past several years exciting fossil and genetic evidence has clarified a key turning point in the early evolution of whales. When researchers compared DNA from whales to that of other living mammals, they concluded that around fifty-five million years ago the earliest whale ancestor split off from the line that evolved into hippos (Figure 6.2).

Two years ago in Pakistan the discovery of a forty-seven-million-year-old fossil skeleton of a proto-whale revealed flipper-like hind feet as well as hoofed forelimbs with ankle bones that flexed like those of modern hoofed mammals like sheep, camels, and hippos. Millions of years after the descendants of the first amphibians crawled onto land, these mammal renegades returned to the sea to become modern whales, dolphins, and manatees. The fossils, genes, and molecular clock give us the smoking-gun evidence.

Two-thirds of my Survey participants accept the fossil and geological evidence for evolution, but their resolve waivers when it comes to the origin of life itself. Nearly half believe that the spark of life was the result of natural chemical reactions, but where do the rest think life came from? Apparently not from outer space. At

Fig. 6.2 Fossil and genetic evidence confirms that whales began to evolve from a hippo-like ancestor (left) around fifty-five million years ago. (Copyright Sinnauer Associates)

least 55% say life didn't come to Earth from other planets. A substantial number (17%) are on the fence about divine creation of the earliest life, and nearly 30% haven't made up their minds one way or the other.

Could it be that these folks are wrestling with the idea of *intelligent* life on other planets? Do they believe that simple life may be possible somewhere out there, but not human complexity? It's hard to get rid of our belief in the special position of humans in the scheme of life. To make matters worse, the hard evidence for human evolution could fit into three steamer trunks.

REFERENCES

Cairns-Smith, A. Graham. 1982. *Genetic Takeover and the Mineral Origins of Life.* Cambridge, U.K.: The Cambridge University Press.

Oparin, Alexander I. 1924 (reprint 1967). "The Origin of Life." *Origin of Life.* J. D. Bernal (ed.). London: Weidenfeld and Nicholson.

Pasteur, Louis. 1864 (reprint 1922). "Lecture to the Sorbonne on Spontaneous Generation." *Oeuvres de Pasteur* 2. S. Pasteur Valley-Padot (ed.). Paris: Masson & Cie.

Peacocke, Arthur. 2000. "The Challenge and Stimulus of the Epic of Evolution to Theology." *Many Worlds: The New Universe, Extraterrestrial Life, and the Theological Implications.* Steven Dick (ed.). Philadelphia: Templeton Foundation Press.

Thomas, Lewis. 1974. *The Lives of a Cell: Notes of a Biology Watcher.* New York: Viking.

WEBSITES

DNA From the Beginning (www.dnaftb.org/dnaftb)

A goldmine of clear, simply put facts that place DNA, genes, and heredity into historical context. Great graphics and links.

Scientific American's "Ask the Expert" (www.sciam.com/askexpert)

This popular magazine's expert site covers nine areas of science. Fascinating assortment of questions with clear, concise answers from a variety of experts. Biology section has current thinking on the origin of life and evolution.

NASA's "Origins" page (http://eis.jpl.nasa.gov/origins)

More than the origin of life on Earth, this very cool site includes info on other solar systems, birth of the Universe, and the search for extraterrestrial life.

SUGGESTED READING

Rare Earth: Why Complex Life Is Uncommon in the Universe, Peter D. Ward and Donald Brownlee, Copernicus, 2000. Two top scientists share their thought provoking argument that complex animal life is rare and we may be all alone in the Universe. Covers all the bases about the origin of life. A must read.

Chapter 7

Three Steamer Trunks

Anthropology is the science which tells us that people are the same the whole world over—except when they are different.
Nancy Banks-Smith

We can pick our friends, but we can't pick our relatives. Yet this kind of wishful thinking resonates in our beliefs about human origins. Our desire to be unique hammers away at our subconscious and creeps into our logic until we succeed in justifying what we want to believe.

Roger D. is a psychologist and counselor in his forties. He has always been intrigued by the human mind, and believes that chemistry is the basis for mental activity—and the key to our humanity.

We know that chemical reactions in the brain play a big role in mental illness and behavior. Animals, such as apes for example, seem to know what will and will not harm the group, but there is no extension of this to others beyond themselves and their immediate group.

Somewhere as we evolved we picked up the idea of what is right or wrong—a conscience. Chemicals came together in our brain to allow for ethics and morality. But nobody knows how and when any of these large changes took place that we associate with being human, because there's no record. Was that where God's finger came in to stir the pot?

In the 1700s Swedish botanist Carl Linnaeus established the Latin system we still use today to give scientific names to plants and animals. Linnaeus christened our species *Homo sapiens sapiens* or "wise human," with the double emphasis on the brainy part shouting "Can you top this?" But hard evidence has forced us to scrap many of our most sacred beliefs about human evolution. For example, we know that our earliest ancestors were small-brained and slept in trees, but also ambled about on two legs. Our large brain was almost an evolutionary afterthought. How did we evolve from tree-dwelling apes into condo-dwelling humans?

Evolutionists tell us that our line originated among the hominoid apes—primates that walked on all fours, lived in tees, and possessed traits that distinguished them from monkeys and other apes. We are a special type of hominoid ape known as a *hominid*—upright and two-legged by definition. Anthropologists use fossil evidence to trace the physical and lifestyle changes that occurred as these apes evolved into modern humans.

Of course it's not really that simple. In the last two decades a windfall of fossil finds have piled up the hard evidence for human evolution. Four new hominid species and a fourth new genus have been uncovered in the last seven years alone. This brings the total number of hominid species to fifteen or seventeen, depending on how you're counting. We know that the life spans of many of these species overlapped for hundreds of thousands of years, suggesting a messy emergence of the human package.

Excitement abounds when anthropologists discover ancient remains that are the oldest. But rapture can turn to rancor if the bones display a mix of primitive and modern traits. These puzzling bones of contention force scientists to rethink evolutionary trees and discard cherished theories. Lines of ancestry are erased and redrawn. Heated arguments break out over claims of heredity. For example, we are uncomfortably drawn to images of the Neandertals. Their heavy brow ridges and short, stocky bodies scream ape. But their expressive faces and penetrating eyes remind us of Aunt Ginny or Uncle Harry. Were they closer to the apes or to us? As a lawyer friend once observed, “When reasonable minds disagree, there simply is not enough evidence.”

Today many scientists depict a concurrent existence of multiple hominid species without connecting the dots. The pattern of human evolution parallels that of all other animals. There were many dead-ends, and the few left standing became us.

By the way, to an evolutionist the word “primitive” means original or earlier, not less. Our primitive ancestors were not less inventive, less creative, or dumb. In their bodies and minds were the original hallmarks of our humanity. At each stage along the trail of human evolution, the survivors were those best adapted both mentally and physically for their time and place.

From Africa to Asia and back across Europe, anthropologists have uncovered the bones of our ancestors. Was evolution’s Garden of Eden in Africa as Darwin predicted? Are we all descended from the first humans to leave Africa and colonize the globe, or is the plot far more complex?

There are some questions that truckloads of fossils may never answer. We share 98.4% of our DNA sequences with those of chimps, our closest primate kin. Recently the Human Genome Project disclosed that humans have a paltry thirty thousand genes—far fewer than the one-hundred thousand originally predicted and not many more than the apes. Chimps don’t play chess or program computers. They also don’t get AIDS, asthma, rheumatoid arthritis, or Alzheimers. But they do make tools, communicate socially, maintain close family bonds, and engage in surprisingly human-like behaviors. A fascinating new study suggests that the differences between chimps and us may be the result of different levels of genetic activity in our brains. Researchers compared the activity of similar genes in various organs of humans and chimps. The results show that levels of activity are comparable in all organs except the brain. Levels of gene activity in the human brain are strikingly higher than those in the brain of chimps. But the products of genes are proteins, not thoughts. How and when did we evolve the complexities of conscience and reason that remain the hallmarks of our species?

We are naive to believe that scientists can completely cut the cord between their culture and their craft. Darwin's own theories were influenced by an orthodox society awash in the competitive fervor of the industrial revolution. But that didn't stop him from claiming in *The Descent of Man* that humans "descended from a hairy, tailed quadruped, probably arboreal in its habits, and an inhabitant of the Old World."

Today we are the only surviving species in our genus—a rarity in the animal kingdom. Over four million years ago two-legged primates took their first steps to becoming modern humans. We are just beginning to locate the signposts of humanity along that trail.

The Gardens of Eden

We left evolution sixty-five million years ago after the demise of the dinosaurs. As continents shifted and climates changed, mammals quickly diversified into vacant niches and fresh territories. The oldest common ancestor of today's primates first appeared in the tropics over eighty million years ago during the reign of the dinosaurs. By fifty-five million years ago these small creatures resembling modern lemurs were well-adapted to life in the treetops. The evolutionary pay-off for primates resided in a handful of distinguishing features that included:

- five grasping fingers or toes at the ends of each limb
- forward-facing eyes and a short snout
- sharp stereoscopic vision
- small litters or one offspring at a time

Primates typically have larger brains that require more energy than other mammals of similar size. For example, the human brain weighs only one-fiftieth of our body, yet consumes up to one-fifth of our energy.

Around fifty million years ago the line of monkeys and apes branched off from the other primates. They had larger rounded heads, two nipples on the chest, and sensitive pads on their fingertips backed by flat nails, not claws. By thirty-four million years ago a group of small hominoid apes had split off from the Old World monkeys. These apes walked on all fours, swung from tree branches, and may have been able to sit up and use their hands to eat. They were the most ancient members of our line.

The arrival of warmer climates around twenty-three million years ago opened up land bridges, prompting a great migration out of Africa. The apes went ape. They diversified rapidly and became as numerous and dominant twenty million years ago as monkeys are today. Fossil apes from this period are rare. But in 1948 Louis Leakey's wife Mary found the first perfectly preserved skull and facial bones of a sixteen-million-year-old hominoid ape.

Many apes had become extinct by ten million years ago when the fossil trail grows cold. The ancient ape that gave rise to both the living great apes and humans remains a mystery. The molecular clock suggests that a common ancestor of humans and chimps split from other apes between five and nine million years ago. During that period intense volcanic activity led to changing climates, dwindling forests, and expanding savannahs. This separated species and encouraged evolution. We know that by six million years ago the two-legged, upright walking hominids had branched off from the chimps.

Despite Darwin's gutsy prediction, no primate fossils were discovered in Africa south of Egypt through the early part of the 20th century. Then in 1924 a young anatomy professor in Johannesburg found a two-million-year-old fossil primate skull in a limestone quarry at Taung in Botswana. The skull of the three to five year old child was no larger than a man's fist and had enclosed a brain the size of an ape's. But the skull also displayed the high rounded shape of a human cranium and suggested an upright posture. This fossil of the "Taung Child" became the first recognized intermediate between apes and humans—the Rosetta Stone of the modern scientific search for human origins.

The Taung Child was initially dismissed as an evolutionary dead-end. Most anthropologists remained mired in the ideological stranglehold of Piltdown and placed humanity's roots just twenty-five miles from Darwin's country home. Meanwhile on the other side of the Atlantic, John Scopes was about to be tried and convicted for teaching evolution.

During the 1930s rich fossil finds of larger-brained hominids from China joined the Taung Child, Java Man, and the Neandertals on a growing list of so-called dead branches on the human family tree.

Finally in 1953 science exposed the Piltdown hoax, vindicated the Taung Child, and confirmed our African roots. The fossils from China and Java were identified as an early human species, but the status of the Neandertals remained in question.

Since then all has not been peaceful in the Gardens of Eden. Louis Leakey, formidable trailblazer in the search for human origins, envisioned a great antiquity for the *Homo* lineage. In 2001 Martin Pickford and Brigitte Senut, excavating in Kenya's rugged Tugen Hills, uncovered parts of a jaw, leg bone, and teeth of what they believed to be the earliest known human ancestor. Christened "Millenium Man," this six-million-year-old chimp-sized hominid pushed pieces of the human package two million years deeper into evolutionary time. Then in 2002 a team led by Michel Brunet excavating in central Africa found a six to seven-million-year-old hominid skull the size of a chimp's. But this "Toumai skull" has facial features similar to those of human ancestors only 1.5 million-years-old. The find is the oldest yet and the first outside of eastern or southern Africa. It demolishes the notion of a single missing link, and emphasizes the role played by evolutionary experimentation in building the bushy human family tree.

Three Steamer Trunks

The scar of the Great Rift slashes across the planet's crust from Turkey southward through Ethiopia and Kenya to reveal fifteen million years of geologic history. Along the East African Rift Valley, preserved in layers of hardened ash from a hundred ancient volcanos, we are unearthing the evidence of our origins. In the Hominid Vault of the Kenya National Museum in Nairobi repose those crown jewels of our ancestors—fossilized bones numbering in the thousands that bear witness to four million years of human evolution.

Hard evidence for human evolution is rare. The chances of finding an early hominid or human fossil are one in ten million. Until recently the entire fossil haul could be jammed into three steamer trunks. Many are mere fragments of bone. Skulls become crushed like china bowls under the weight of rocks and time, and few skeletons remain intact. Most fossils are teeth, the virtually indestructible parts of our body. Fourteen bones from the face and eight bones of the cranial vault, or about one tenth of the total bones in our body, are key to identifying an early hominid. When we add teeth, limb bones, and the pelvis, a true portrait of our evolutionary ancestor emerges.

Anthropologists must draw conclusions about our past from sparse evidence. Each new fossil is so rare that it may vary significantly from all others. The normal range of variation within a species becomes apparent only after examining fossils from a large number of individuals. A sample of one doesn't give us an accurate picture of this variation and can lead to misidentification of new finds. For example, if aliens from outer space landed in New York City, their abduction of one person would give them a very biased sample of modern humans. We are all the same, but also very different.

In the politics of paleontology, there are the "lumpers" and the "splitters." Splitters are more often inclined to award a new species name to a fossil based on unique variations than the lumpers, who hesitate to add another category to an already crowded field. As fossil hunters dig up more bones, they substantiate the extent of variation in brain size, tooth pattern, skull shape, and other traits among our early ancestors. Existing species are consolidated and old names fall by the wayside. We draw our remote ancestors closer to us with nicknames like "Handy Man" and "Lucy." Who were these guys and gals? What can their fossils tell us about the roots of our humanity?

Walking the Walk

In 1974 in the Afar Triangle of Ethiopia, Donald Johanson found a three-million-year-old hominid skeleton he named "Lucy." This small-brained creature was a new species in the Southern Ape family and quickly supplanted the Taung child as the oldest known human ancestor.

The skeleton of "Lucy" was headless, but a few cranial and jaw fragments suggested that her brain was the size of a chimp's. At three and a half feet tall she was fully adult. Above the neck Lucy was more ape than human. Her long arms and hands hinted she may have lived or slept mostly in trees. But down below, her pelvis and limb bones revealed us that she had walked fully upright like us.

Two years later Johanson found "Lucy's Family"—a group of several individuals who came in two distinct sizes. This discovery stirred debate over the possibility that the fossils represented two separate species living together. In 1993 the mystery was solved when discovery of fossils of another large male demonstrated that extreme size differences between the sexes was common in "Lucy's" species. Some males were 50% larger than the average female. Today we see this phenomenon of size disparity in male and female gorillas. As hominids evolved, this size gap narrowed, and for the past 1.8 million years has been fairly constant in humans. Size differences between the sexes are just one example of the danger of drawing broad conclusions based on small or incomplete samples.

In 1978 a team led by Mary Leakey found the 3.5 million-year-old footprints of Lucy's ancestors preserved in hardened volcanic ash. These early hominids strode confidently across the dry dust of Tanzania half a million years before Lucy was born. In 1995 another Leakey team found the 4.1 million-year-old fossil of a new small-brained hominid species who was walking upright a million years before Lucy. Cultural bias has always maintained a brain-centered focus on human evolution. But the evidence clearly shows that an upright posture led the way.

Tool Time in Tanzania

The oldest stone tools are crudely chipped cobbles found by Louis Leakey in the 1930s while excavating in the Olduvai Gorge in Tanzania. These simple "Olduwan" tools date as far back as two and a half million years ago. Thirty years later Leakey found the skull of a larger brained hominid that he believed was closely related to humans. He named the creature "Handy Man" because he thought he had found the maker of the Olduwan stone tools. Handy Man became the oldest member of the human genus *Homo*.

For many years anthropologists considered toolmaking one of the defining hallmarks of humanity. We now recognize that many animals, including chimpanzees, orangutans, and birds, use natural objects as tools. A few minutes watching chimps peel fruit or use a twig to probe for juicy insects should do away with the myth of the mighty human thumb.

Chimps will occasionally toss rocks underhanded, but "Lucy's" fingers could rotate slightly inward and secure a finer grip on small objects. With practice she could have hurled a rock, or even a baseball, with some degree of accuracy. Her upright posture would have added leverage and balance to her throw. But toolmaking requires the intelligence to visualize the finished product as well as the manual dexterity to produce it. The first tools were probably naturally sharpened stones or crude wood implements. But in the early humans like "Handy Man," we first observe a little finger that can cross the palm, creating the manipulative grasp essential for chipping sharp edges on stone. It's really our opposable pinkie that sets us apart from the apes.

What kicked off the evolution of the human line and its split from the family of "Lucy" and the Taung Child? Major evolutionary events correlate with major ecological change. In his book, *Children of the Ice Age*, paleobiologist Steven Stanley suggests that the beginning of the Ice Age three million years ago triggered the evolution of the first humans. Africa suddenly got cooler, windier, and drier just prior to the first evidence of stone tools and emergence of the human genus *Homo* around 2.8 million years ago. Open forests shrank and grasslands expanded. Evolution would have favored hominids who were more nimble on two legs and had the intelligence to defend themselves out in the open.

Man-[and Woman]-the-Scavenger

When Louis Leakey found the Olduwan stone tools, he was really searching for Man-the-Intelligent-Hunter. For many years anthropologists based their ideas of early human subsistence on modern hunter-gatherer groups. But soon archeological evidence from ancient campsites and cut-marks on fossilized animal bones, replaced the idea of Man-the-Hunter with Man-[and Woman]-the-Scavenger.

Scavenging is smart. The meal is already dead so it's a sure thing. Scavenging saves time and energy, and avoids injury. Many modern mammals such as hyenas, lions, and leopards are both hunters and scavengers. But scavenging is risky when large predators defend their kill, so scavengers must develop strategies to both locate and protect dinner.

Meat was a rich source of protein energy needed to power the larger brain of *Homo erectus*—an early human species that appeared around two million years ago. With bigger brains came cooperation in hunting. Small game could be caught in snares or by throwing stones, and larger game was driven over cliffs. Soon the dawning recognition that making a tool one day meant hunting success the next spurred invention and creativity.

The First Family

Brain size remained stable from the time of *Homo erectus* two million years ago until six-hundred thousand years ago when we see pronounced brain growth relative to body size. By three-hundred thousand years ago brain size began to approach that of modern humans. At the same time the pace of technological innovation picked up.

But bigger brains carry a higher evolutionary price tag. Human babies are born with an undeveloped brain and soft skull to fit through the birth canal. They remain helpless during the first year's surge in brain growth. Many anthropologists speculate that this burden of infant care prompted male-female bonding among our earliest ancestors. A male would have gained personal power by providing for one or more children whom he had sired. The evolutionary advantage was enhanced survival of the group through strength in numbers.

We Were Not Alone

During the 1970s and 1980s evidence began to emerge that would shatter the image of a single, straightforward line to modern humans. Not one but several hominid species had trudged together along the trail of human evolution. Copious speciation was the rule in human evolution just as it was in the evolution of other animals. The precise way in which several early hominid species resolved into a single human species may never be known. Barring an ancient polaroid snapshot of the event, we will never know whether the Taung Child's descendants considered Handy Man a "hunk" or an animal. We may think that Neandertals look a lot like us, but to early modern humans they may have resembled the Abominable Snowman—and smelled even worse! A solitary "missing link" and a straightforward dotted line may be the greatest myths of human evolution. The modern human genome is the true survivor in our story.

Most evolutionists now believe that sometime after three million years ago Lucy's line split into two branches, one leading to big-brained *Homo erectus*, and the other to more robust ape-like hominids. Several species of robust hominids lingered alongside early humans until one million years ago when a cold snap spelled their ruin and cleared the playing field for the emergence of modern humans. Today one of the most passionate debates in human evolution revolves around the fate of the Neandertals. Were they our brothers and sisters, or only distant cousins?

The Great Cave Man Controversy

In 1856 the bones of an ancient human were found in a Cave in the Neander Valley near Dusseldorf, Germany. His heavy browridges, bizarre build, and the largest nasal and sinus cavities ever seen made him an immediate curiosity. But his skull suggested a modern brain and he looked hauntingly like us (Figure 7.1).

Fig. 7.1 Reconstruction of a three-and-a-half-year-old Neandertal child from remains recovered in 1926 from the Devil's Tower site in Gibraltar. (Copyright M. Ponce de Leon and Ch. Zollikofer, MultiMedia Lab, University of Zurich/Switzerland)

Initially anthropologists claimed that the bones belonged to a north European barbarian. Noting the short, bowed legs, they presumed the fellow had done a lot of riding and might have been a Cossack soldier left stranded in the cave. Some suspected his skeletal deformities had been caused by a case of childhood rickets. As more Neandertal skeletons turned up, puzzled scientists tried to fit them into the human race.

In the 1800s Europe was the economic, political, and scientific powerhouse of Western Civilization, but paleoanthropology was in its infancy. Biblical literacy and the Genesis account of a singular Creation dominated belief in human origins. Any thought that Neandertals might be a different type of human being didn't even pop up on the radar screen. Most intellects of the time agreed that if humans had existed before Noah's Flood, they must have been Europeans. As a result the peculiar anatomy of Neandertals could only be attributed to disease, trauma, or congenital defect, and associated with mental deficiency. Neandertals quickly assumed the role of intellectually flawed, morally deficit cave dwellers.

In 1864 geologist William King first publicly acknowledged that another human besides *Homo sapiens* had existed. While studying a Neandertal skull, King had been struck by the absence of shapes normally found in the human forehead and face. He remarked that the "thoughts and desires that once dwelt within it never soared beyond those of the brute," and reluctantly branded Neandertals a separate but inferior brand of humans.

Four years later French paleontologist Eduoard Lartet and the English banker Henry Christy were excavating a small rock shelter near Les Eyzies in the south of France when they found five burials of apparently modern humans. This wasn't the first discovery of its kind, but it was to become the most famous. These prehistoric people, named Cro-Magnons or "Big Cliff" for the site of their discovery, were the first modern humans in Europe.

As the hard evidence for human evolution accumulated, novel theories sprouted like dandelions. All new finds were forced into a single line of human descent. Most scientists believed that the Cro-Magnons arrived long after the Neandertals. Java Man was thought to have been a forerunner of the Neandertals. When further excavations revealed that Java Man and Neandertal were roughly of the same age, the obvious dilemma was ignored.

Between 1908 and 1911 four cave sites in the south of France yielded multiple skeletons of Neandertals. When added to earlier finds these fossils defined the broad range of Neandertal anatomy. Marcellin Boule, an influential French paleontologist, set the tone by insisting that Neandertals were an ape-like offshoot of the human line that died out without a trace. But the scientific caricature of Neandertals as dim-witted brutes was finally put to rest in the 1950s when study of the very skeleton Boule had used as the standard by which to judge all Neandertals was shown to not differ significantly from normal variations found in modern human populations. It seems the poor fellow had simply suffered from a bad case of osteoarthritis and old age. Were Neandertals simply early modern humans with an overbite?

In the last fifty years we have acquired a remarkable picture of Neandertals as:

- Compassionate Care Givers—Life was brutal for the Neandertals. Half of all Neandertal fossils are children under the age of eleven. But a recent computer reconstruction of the skull of a Neandertal youth shows he had been hit over the head with a sharp weapon and had survived the violent attack only by being nursed back to health.
- Worshipers—Evidence for religious belief among Neandertals is scarce. However ritualized arrangements of cave bear skulls at sites in Switzerland and France suggest some form of cult devotion. Discovery of pollen grains from flowers beneath a Neandertal skeleton in a cave in Shanidar, Iraq suggests an intentional burial, perhaps accompanied by simple ritual. But we'll never know whether Neandertals believed in an afterlife.
- Cannibals—There is substantial evidence that Neandertals stripped the meat off human bones and

extracted the marrow, either as a source of nourishment or as a ritualistic practice.

- Artists—Neandertals scratched faint patterns on bone and rock, and polished animal teeth to wear as pendants.
- Providers—Neandertal hunters used short, stone-tipped spears to wound prey at close range. They also trapped small animals, drove larger game over cliffs, and probably relied heavily on scavenging. But there is little evidence that they followed migrating herds or exploited other protein-rich resources such as salmon found in northern Europe.
- Architects—Neandertals literally lived hand-to-mouth, cave-to-cave. But evidence of their temporary camps in open areas and rudimentary huts assembled from mammoth bones reveals the ancient roots of architecture.

Could Neandertals speak? In the 1960s the discovery of a Neandertal hyoid, the throat bone that permits speech in humans, created speculation that its owner could talk. But the position of the Neandertal tongue, and the size and shape of their nasal cavities, may have prevented formation of some vowels and consonants. Neandertals could certainly communicate, but they probably sounded like they had a mouthful of food and a bad head cold. Whether they had language based on grammar and syntax will never be known.

Neandertals flourished in Europe for over two-hundred thousand years during the worst of the Ice Age. They were a highly successful, cold-adapted derivative of *Homo erectus*. But as warmer climates returned to Europe, Neandertal anatomy and technology became a disadvantage and their populations dwindled. Small groups survived in cooler areas throughout Europe where they lived alongside modern Cro-Magnons for perhaps ten thousand years. The last of the Neandertals retreated to a stronghold in the mountains of southern Spain, where they died out twenty-seven thousand years ago.

There is scant evidence for Neandertal-Cro Magnon interaction. A rash of Neandertal finds throughout Europe and the Middle East have raised many questions. In some cases the bones were accompanied by more technologically advanced tools such as handaxes and backed knives similar to those made by early modern humans. But it's impossible to say whether these are evidence of invasion or innovation. Most remains exhibit classic Neandertal traits while others contain a tantalizing mix of ancient and modern features. Was it possible for Neandertals and fully modern humans to share hunting grounds but not genes for thousands of years?

Today the majority of anthropologists and evolutionists adhere to the "Out-of-Africa" or replacement theory. They believe that fully modern humans began to evolve in a single place, namely Africa, beginning around two-hundred thousand years ago. As they evolved they migrated in multiple waves out of Africa and into the Middle East, Asia, and Europe, replacing all less advanced human populations with little or no interbreeding. Neandertals belong to a separate human group and are our distant cousins at best.

But not everyone is convinced. The opposing camp, led by anthropologist Milford Wolpoff, claims that *Homo sapiens* arose independently in multiple regions of the world. They insist that indigenous populations of *Homo erectus* throughout Europe, Asia, and the Middle East continued to evolve for two million years as genetically distinct races of a single interbreeding species. Migration and interbreeding mixed their genes with other evolving humans until the major features of anatomically modern humans converged into a single package and a single species—*Homo sapiens*. According to Wolpoff and fellow regionalists, Neandertals could also have left their stamp on the modern human genome despite their apparently swift replacement by Cro-Magnons in Western Europe.

In the mid 1970s a team of biochemists studying the genetic relationship among modern ethnic groups stunned anthropology and nearly stopped the regionalists in their tracks. Using a DNA clock based on a small number of genes in mitochondria (cellular powerhouses uniquely inherited only from our mothers), researchers concluded that all modern humans share a common female ancestor who lived two-hundred thousand years ago in Africa. This African Eve wasn't one woman, but a small population of closely related females whose mitochondrial genes survive in each of us. Additional research using this same technique has recently confirmed the multiple migrations of increasingly advanced humans out of Africa and into Eurasia. And in 1997 scientists isolated DNA from the thirty-thousand-year-old bones of a Neandertal and compared it to modern human DNA. Their conclusion? Neandertals were not our ancestors and did not interbreed with modern humans.

Case closed? Not so fast. Just to keep everyone on their toes, finer DNA analyses now hint at the possibility that small amounts of Neandertal DNA may have survived in our genes. The jury is still out.

On the Road Again

The prize for finding the most hominid fossils doesn't go to someone named Leakey or Johanson, but to Kimoya Kimeu—a native Kenyan and Leakey-trained field observer.

In 1984 Kimeu topped off an already impressive career by finding the most complete skeleton known of an early human dating to 1.6 million years ago. The teenager nicknamed "Nariokotome Boy" had an unusually tall and slender body build that suited him for life out on the hot, open savannah. His larger brain would have made him an astute scavenger, and his beautiful stone handaxes flaked on both sides indicated that their maker had used a "mental template" to conceive and sculpt a standard shape.

"Cash, Jordan. That's what separates man from the apes."

Fig. 7.2 ©The New Yorker Collection 1993 Leo Cullum from cartoonbank.com. All Rights Reserved.

We left human evolution on the plains of Africa two million years ago when *Homo erectus* showed up with a larger brain and a new tool kit that confirmed higher intelligence. These guys and gals were the first world travelers, heading out of Africa as early as 1.8 million years ago. Where they went next is still unclear. Here's what we know.

- *Homo erectus* left Africa by 1.8 million years ago. Some of the most ancient undisputed human fossils outside Africa are two skulls found in the Soviet Republic of Georgia that date to 1.7 million years ago. These early travelers were already one-third of the way to China. Small groups of *Homo erectus* reached Asia where their descendants acquired nicknames like Java Man and Peking Man. *Homo erectus* survived in Asia until thirty to forty thousand years ago, maintaining for nearly a million years the same simple tool kit and life style they had brought with them from Africa.

- Meanwhile back in the Garden of Eden, after *Homo erectus* hit the road to the Far East, her African descendants kept on evolving. As early as one million years ago the modern human skull began to take shape. Around that same time a second migratory wave of our early ancestors left Africa for Europe.

Until recently no fossils of early humans dating from a half a million to a little over a million years ago had been found in Europe. Then in 1994 paleontologists excavating a cave in northern Spain found the fossil remains of six early humans amidst a scene of carnage and cannibalism. The eight-hundred-thousand-year-old bones of an eleven year-old boy and five adults had been filleted by a sharply flaked stone tool and then eaten. The boy's face displayed a startling mosaic of primitive and modern traits never seen before. Many anthropologists now believe these folks were descendants of that second wave of emigrants from Africa, and the great-grandparents of the Neandertals who appeared six-hundred thousand years later.

- Back in Africa humans were still evolving, and by two-hundred thousand years ago they looked very modern. The earliest confirmed evidence for anatomically modern humans in Africa dates to one-hundred-twenty-five thousand years ago, but there is evidence that by one-hundred-thirty thousand years ago fully modern humans had already arrived in the Middle East. Encouraged by a warming trend, more left Africa and reached Israel by ninety-two thousand years ago. There they lived alongside more primitive *Homo erectus* for at least fifty thousand years. By forty thousand years ago fully modern humans showed up in Europe as the Cro-Magnons.

Throughout most of human history the world population has been unbelievably small. In the heyday of the Neandertals, perhaps no more than a hundred thousand of them inhabited Europe and western Asia. Early humans lived in small bands of eight to twelve individuals who came in contact only occasionally, perhaps to find a mate. On the extreme edges of human evolution, these small populations quickly diverged as their genetic differences became concentrated and amplified. Exciting new finds corroborate multiple waves of immigration, but scarce fossils prevent tracking the movement of every small group of early modern humans. Some of these travelers probably got homesick and returned to Africa, planting new genetic seeds in the Garden of Eden and carrying innovative technologies across continents.

Four Million Years of Humanity

Clothes really don't make the man or woman, and the fossilized bones of our ancestors don't make us human. What does?

Frank C. retired after twenty-five years as a policeman. He has seen human nature at it's best and worst—an experience that clearly colors his belief about what makes us human.

We evolved, but when did we start changing into being human? It was such a gradual thing we may never know when it began. But there are certain marks, like communication, language, religion, a learned moral code, and some feeling of the past and the future. Chimps and other animals live in the moment. We have a long way to go in evolution. I think we're still evolving. We're not complete.

Just like Frank, many of us prefer to keep the links to our animal origins tenuous and brief. But with the ascendancy of researchers like Jane Goodall and her studies of chimps in the Gombe

Forest has come new appreciation for animal behavior when viewed through the lens of evolution. Chimps appear to fill both a behavioral and genetic gap between humans and other animals. Nevertheless, something is missing.

American archaeologist Alexander Marshack coined the phrase "the human capacity" to describe this elusive quality of behavior and psychology we see for the first time in the Ice Age art and symbolism of Cro-Magnons. Only in modern humans do we find the unique combination and synergy of elements that form the human capacity, including:

- Consciousness—Philosopher Thomas Nagel once asked, "What is it like to be a bat?" Only the bat knows. And if it does, then we can say that the bat is conscious. Do animals think? Are they consciously aware? We may never know.
- Morality—Most people believe that the crux of our humanity is our *consciousness with a conscience*. We lead human, not ape-like, social lives. Much of our additional brain power is used to keep us going psychologically and socially as moral human beings. Despite anecdotal evidence for altruism and ethics in animals, we may never know if animals behave the way they do because they think about the consequences.

In the *Descent of Man*, Darwin wrote,

A moral being is one who is capable of reflecting on his past actions and their motives—of approving of some and disapproving of others; and the fact that man is the one being who deserves this designation, is the greatest of all distinctions between him and the lower animals.

- Intelligence—In his book *Wild Minds: What Animals Really Think*, Harvard psychologist Marc Hauser claims that all animals are equipped with the basic mental tools to solve specific problems. Evolution doesn't always favor a higher intelligence, but it has in the case of humans.

The human brain is the one organ where size really does matter. But not just any size. During the last three million years our brain has tripled in size compared with a chimp's. Most of that increase has been in the cortex, a thin rind of convoluted gray matter that surrounds the more primitive mid-brain. In this uniquely enhanced portion of our brain reside the logic circuits and working memory of the human capacity.

A recent discovery has uncovered differences between the brains of chimps and humans that links physical evolution, communication, and intelligence. Casts of chimp brains were compared to those taken from the fossil skulls of evolving humans from hominids like "Lucy" through Neandertals and early modern humans. Computerized imaging of the casts reveal significant expansion of the right side of the evolving human brain compared to that of chimps. The greatest degree of bulging and asymmetry occurs in the brains of Neandertals and early modern humans. Comparison of brain *activity* in chimps and modern humans also differs in five "hot spots." Three of these spots are the same areas that shown differential physical growth in humans during two and a half million years of evolution. Researchers believe that the expanded hot spots in our brain aid in decoding the rhythm, tone, and emotion of speech, all keys to intelligent communication.

- Sense of Self—Only dolphins, chimps, and humans are known to react to their image in a mirror. Psychologists believe this is an important first step in building an awareness that we are individuals apart from our surroundings. But as humans we also sense ourselves in relationship to others. We develop an acute awareness of what others might be thinking, feeling, or planning. This is a uniquely human survival advantage called social intelligence.
- Social Behavior—Humans are instinctively social animals, just like bees, birds, and the chimps of the

Mihale Mountains of Tanzania who raise their arms and clasp palms in a "high five" to reinforce identity and recognize rank in the group. But humans have survived and flourished by taking social interaction to new heights. Early in our evolution we learned that the survival strength of a group is greater than the sum of its individual members. The story of human evolution is the story of strength in numbers optimized through social intelligence and morality.

- Culture—In 1999 a landmark study of animal behavior concluded that chimps have culture. The researchers identified over thirty-five behavior traits that vary with the group, including grooming, eating, and grabbing the attention of other chimps. The cultural high water mark in human evolution occurred thirty-five thousand years ago. A decisive change in culture and behavior known as the Great Leap Forward appeared with the arrival of the Cro-Magnons. These folks didn't just look different, they *thought* different. They were innovative. Unlike Neandertals, their tools were obvious in function, diverse in form, and often combined stone, bone, and wood. They fished with barbed harpoons and carried on long distance trade for raw materials. Cro-Magnons were the Michelangelos of the Pleistocene. Their aesthetic sense shines in ivory and bone carvings, stone sculpture, and naturalistic cave paintings. Cro-Magnons had style.

Our ancestors were anatomically modern but culturally primitive for over fifty thousand years before the Cro-Magnons appeared. What was the source of our Great Leap Forward? Anthropologist Ian Tattersall of the American Museum of Natural History believes that the human capacity was a by-product of other evolutionary changes, waiting to be unleashed by a cultural stimulus. That stimulus was *language*.

- Language—Only humans communicate ideas through complex language and grammar. This drives invention, novel connections, and new ideas—like writing this book. By the time *Homo sapiens* appeared, most of the anatomical prerequisites for speech had been around for several hundred thousand years. Perhaps a fractional genetic change between earlier humans and fully modern Cro-Magnons was enough to alter the vocal tract in a way that made complex speech possible.

Suddenly and without warning, a profusion of unique behaviors are detected in the Cro-Magnons. For the first time we can infer from hard evidence the presence of symbolic thought and elements of the human capacity. Language and symbolic thought go hand-in-hand. Once language evolved to support innovation, the Great Leap Forward was possible. We were no longer dependent on our genes.

Evolution of the human capacity, like our physical evolution, is a story of experimentation and eventful complexity. We have continued to build the human capacity through social interaction, accumulation of new knowledge, and expanded communication.

Elaine K. is a retired secretary in her sixties. Though busy being a mother, grandmother, homemaker, and churchgoer, she still has enough time on her hands to read about her favorite topic, human evolution.

I've read a couple books by Richard Leakey, and one about "Lucy" and her family. And I have read the newspaper and magazine stories about how we're all descended from a woman in Africa they call "Eve."

I wonder about the people who make human evolution their life's work. They seem to have so little evidence to go on—a hank of hair and a piece of bone. They might have a jaw with a few teeth in it and they build a human out of it. In one sense it's fascinating to read about, but on the other hand it's amusing. But I still buy into it. The environment was right and we grew like Topsy!

A majority of my Survey participants agree with Elaine that modern humans evolved and have been around for at least forty thousand years.

Next let's turn to the question of the soul, and the part it plays in our beliefs about human origins.

REFERENCES

Hauser, Marc D. 2000. *Wild Minds: What Animals Really Think*. New York: Henry Holt.

Stanley, Steven M. 1998. *Children of the Ice Age: How a Global Catastrophe Allowed Humans to Evolve*. New York: W. H. Freeman.

Tattersall, Ian. 1995. *The Fossil Trail: How We Know What We Think We Know About Human Evolution*. New York: Oxford University Press.

Trinkhaus, Erik and Pat Shipman. 1993. *The Neandertals*. New York: Knopf.

WEBSITES

Becoming Human (www.becominghuman.org)

Outstanding website from Donald Johanson's Institute for Human Origins at Arizona State. Puts you in the middle of the action in the search for human origins. Includes a Web documentary, up-to-date info on latest finds and on-site field activities. Great graphics and links.

Neanderthals and Modern Humans: A Regional Guide (www.neanderthal-modern.com)

Well-referenced site presents an even-handed regionalist perspective on the hard evidence from Europe and Asia about those low-browed hominids and their relationship to us. Up-to date information with excellent links to books, museums, societies, and research institutions.

Neandertals: A Cyber Perspective (http://thunder.indstate.edu/~ramanank/index2.html)

Award-winning site tells you everything you wanted to know about life among the Neandertals. Written for a general audience, well-researched, educational.

SUGGESTED READING

Ian Tattersall, *Becoming Human: Evolution and Human Uniqueness*, Harcourt Brace, New York, 1998. The fossil trail and the search for our humanity.

Erik Trinkhaus and Pat Shipman, *The Neandertals*, Knopf, New York, 1993. History, science, and entertainment for anyone interested in human origins.

Alan Walker and Pat Shipman, *The Wisdom of the Bones: In Search of Human Origins*, Vintage Books, 1997. Human evolution from the man who dug up Nariokotome Boy.

Milford Wolpoff and Rachel Carson, *Race and Human Evolution: A Fatal Attraction*, Westview Press, 1998. The multiregional theory of the evolution of modern humans.

Chapter 8

Where East Meets West

We do not believe in immortality because we can prove it, but we try to prove it because we cannot help believing it.

Harriet Martineau, 19th C. Journalist

Mona K. is the middle-aged widow of an American diplomat, and has lived most of her adult life in Europe and Asia. She trusts science and believes that the alleged healing power of prayer is an illusion that has a rational explanation. She rejects Adam and Eve and accepts the theory of evolution as the most plausible explanation for human origins. But Mona's certainty falters when it comes to the question of life itself.

Evolution doesn't explain how life began. What put the spark of intelligence in us that makes us different from the animals? Some people think it was all chemistry like electricity. When you're pregnant it really does feel like a spark. I guess this is what we call the soul.

I would like to know if anything happens to the soul when we leave the body. The Hindus believe that it joins the stream of life, the Great River, and has continuity. It's part of our search for immortality.

Mona acknowledges the existence of a *soul*—that elusive ingredient in the human package that may survive death. She is not alone. Over 75% of Americans believe in some form of life after death, and half the world's population believes in reincarnation.[1] For these folks, the answer to where we came from lies not in the origin of our mortal bones, but in the evolution of our immortal souls.

Most of us acknowledge a place deep within us where the tangible and intangible connect. We search for soul mates and say that music has the power to move us body and soul. And many of us sense that, as the Greek poet Pindar rhymed, "All human bodies yield to Death's decree. The soul survives to all eternity."

During the Pyramid Age, ritual entombment of Egyptian Pharaohs reenacted the rebirth of the divine soul of the king. Tomb and coffin decorations depicted the human soul as a bird that could escape the darkness of the tomb, fly through the underworld with the sun, then return to comfort the body of the deceased (see Figure 8.1).

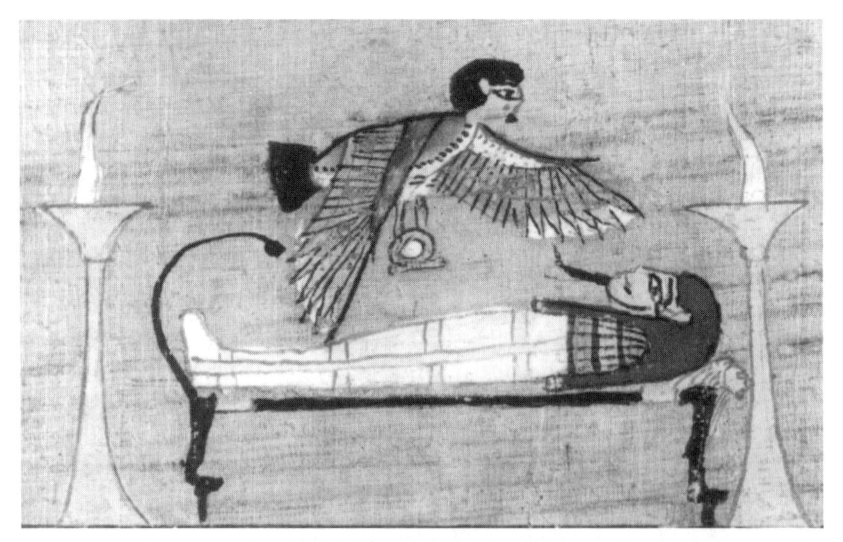

Fig. 8.1 This vignette from the Egyptian Book of the Dead of Ani depicts the soul as a human-headed falcon that reunites with the mummy of the deceased. © Copyright The British Museum

But after 2250 B.C. a political breakdown of the Old Kingdom and collapse in the social order of the Pyramid Age sparked a period of dynastic disarray, intellectual fervor, and religious skepticism. For the first time in recorded history, state religion became universally accessible to commoners, and fundamental questions were raised about the nature of the soul and its survival after death.

The problem of a soul is twofold. Does a soul exist? If so, what is it made of? Most early philosophers and naturalists conceded that the soul, whether material or immaterial, can't be known through direct observation. It can only be inferred, just as we must infer the unseen atomic structure of objects.

Plato wrote that the soul organizes and energizes the body in the same way that the "world soul" of the early Greeks moved the Universe. It is that part of us that may appear to die, but is born again. But Plato cautioned that "to attain any assured knowledge about the soul is one of the most difficult things in the world."

As expected, Aristotle disagreed. He argued that the soul is real, inseparable from the body, and should be a prime target for investigation. He concluded,

> *What is soul? It is substance in the sense which corresponds to the definitive formula of a thing's essence. That means it is 'the essential whatness' of a body.*

Now that really cleared things up! Aristotle's murky musings on the soul have kept philosophers in business down through the ages.

During the 1600s French philosopher and mathematician René Descartes set out on a nine-year journey in search of Truth. His cloistered Jesuit education had given him a moral code but left him doubting all knowledge except the reasoned proofs of his beloved geometry. Descartes roamed Europe, observing everything through his lens of doubt. In a famous feat of negative logic, Descartes concluded that by doubting the existence of everything, he could not doubt that he was doubting. Or in his own words,

> *This truth,'Cogito ergo sum' [I think or I know, therefore I am], was so certain . . . that the skeptics were incapable of shaking it. And then, examining attentively that which I was, I saw . . . that this "me," the soul by which I am what I am, is entirely distinct from body.*

Descartes' circular proof defined the mind as separate and disconnected from the body. This created a dilemma that looms large each time we try to analyze the nature of the soul or consciousness—it's *ourselves* studying *our self.*

Theological differences aside, any frank discussion of where we came from must reckon with the soul. Along with evolution, creationism, and life from other planets, most people include a human soul in the origins puzzle. Whether we embrace it wholeheartedly or reject it outright, the soul intrudes upon our belief because it satisfies our universal longing for:

- immortality (soul as ultimate survival tool)
- uniqueness (soul as basis for human superiority)

Today science is searching for Descartes' soul in the chemistry of the brain. Is a soul simply organized static thrown off by the electrical impulses of neurons and chemical transmitters? Over two thousand years ago the Roman poet Lucretius, the soul maverick among his peers, outlined the modern scientific approach to the soul when he wrote:

- *The mind which we often call the understanding is no less part of the man than hand and foot and eyes are parts of the whole living creature.* (The mind or consciousness is physical in nature)
- *The mind and the soul are kept together in close union and make up a single nature.* (Consciousness and "soul" are the same)
- *This same principle teaches that the nature of the mind and soul is bodily and formed of exceedingly minute bodies.* (Both consciousness and "soul" are tangible and made of atoms)
- *When the enveloping body has been all broken up and the vital airs have been forced out, you must admit that the senses of the mind and the soul are dissolved, since the cause of destruction is one and inseparable for both body and soul.* (When we die the soul dies with us)

Lucretius not only affirmed the soul's materiality but also rejected immortality and reincarnation, arguing, "If the nature of the soul is immortal and makes its way into our body at the time of birth, why do we retain no traces of past actions?" But some people claim they do exactly that.

Soul Brothers and Sisters

Modern life has torn the web that connects us with nature. Many people have experienced this violation as a sense of personal spiritual loss. In the past decade we've seen a marked rise in spirituality among Americans yearning to fill this void. From the born-again faithful to avid environmentalists and New Agers, we are reconnecting by worshipping in crystal cathedrals, preserving primal landscapes, and recapturing past lives. But long before Shirley MacLaine regaled us with stories of her past life as a jester in the court of Louis XV, and Sylvester Stallone revealed to us that he had once lived as a monkey in Guatemala, there was the curious case of Virginia Tighe.

From 1952 to 1953 Virginia participated in six hypnotic sessions conducted by Morey Bernstein, a Colorado businessman, amateur hypnotist, and believer in reincarnation. The twenty-nine-year-old housewife was an excellent subject and quickly fell into a deep hypnotic trance. Bernstein taped the sessions which were attended by several witnesses, including Virginia's husband.

During the first session Bernstein asked Virginia to think back to a time before her birth and describe any scenes of faraway places that came to mind. The Chicago native suddenly started speaking in an Irish brogue and identified herself as Bridey Murphy, an Irish lass born in Cork in 1798. She described her childless marriage to a lawyer named Sean Brian Joseph MacCarthy, and their rather uneventful life in Belfast. "Bridey" remembered suffering a crippling fall in her early sixties and remaining in frail health until her death in 1864 at the age of sixty-six.

Bernstein and the other witnesses were convinced that Virginia's answers were spontaneous. Her thick brogue, use of strange sounding Irish words, and detailed knowledge of life in

Ireland during the 1800s seemed authentic. At one point Bernstein gave his subject a posthypnotic suggestion to dance an Irish jig. Upon awakening, and after a little coaxing from Bernstein, Virginia began to move her feet rapidly in little dance steps. This was especially compelling because those who knew Virginia said she was such a poor dancer that she couldn't have danced a jig if her life had depended on it.

Neither Virginia nor her husband believed in reincarnation before the sessions. Afterwards they said it seemed the only explanation. According to Virginia's husband, the couple didn't own any reference books, encyclopedias, or even a library card, saying "I know all that information could not be pouring out of Virginia."

In 1954 the *Denver Post* published several articles about "Bridey Murphy," defending reincarnation as the explanation for Virginia's memories. Reader response was so positive that Doubleday hired Bernstein to write a book about the case. *The Search for Bridey Murphy*, published in 1956, became an instant best-seller.

"Bridey" mania swept the country, stimulating interest and research into the source of knowledge about past lives. Books on reincarnation sold like hotcakes. Hypnotic regression became the rage. Invitations to Bridey Murphy costume parties read "Come as you were." Bernstein's book went into newspaper syndication and movie rights were sold. A long-playing record of the first Bridey Murphy hypnotic session flew off store shelves.

But at the peak of the "Bridey" frenzy, investigative reporters from the *Chicago American*, a Hearst rival of the *Denver Post*, wrote a scathing exposé of Virginia's early life in Chicago. They claimed that her recollections were childhood memories, not past life revelations. Reporters said they had found the real "Bridey Murphy"—a Mrs. Bridie Murphy Corkell from County Mayo, Ireland, who had been Virginia's neighbor across the street in Chicago. A local high school teacher told the reporters that Virginia had been active in dramatics and was known for delivering Irish monologues in a heavy brogue. They also claimed to have discovered that Virginia's favorite Irish aunt, long since deceased, had often told stories about life in Ireland. This damning glut of facts and details from Virginia's real life so closely paralleled her hypnotic utterances that it destroyed all credibility for her Bridey Murphy story.

The *Denver Post* countered by sending their best reporter, a skeptic of reincarnation, to verify the *American's* story. He discovered that facts had been manipulated or completely invented by the Hearst reporters. He was also able to validate most of Virginia's side of the story, but by then the damage had been done.

A subsequent review of Bernstein's tapes revealed he had generously embellished Virginia's recollections in his book. Attempts to verify names, dates, and locales from Virginia's memories of life in Ireland yielded no evidence to corroborate that Bridey Murphy or her husband had ever existed.

Reincarnation is a belief held in every known civilization. The Greek historian Herodotus reported that,

> *The Egyptians were the first to broach the opinion that the soul of man is immortal, and that, when the body dies, it enters into the form of an animal which is born at the moment, thence passing on from one animal into another, until it has cycled through the forms of all the creatures which tenant the Earth, the water, and the air, after which it enters again into a human frame, and is born anew.*

Hindus believe that individual souls ascend through lower forms of life until they reach a human body. There the automatic reincarnation stops. The soul gains self-consciousness and the law of *karma* kicks in—which simply put means that what you do in this life (cause) determines the type of life you will experience in your next reincarnation (effect). The ultimate goal of this endless transmigration of souls from one mortal to the next is a state of selflessness that Buddhists call Nirvana.

Reincarnation is most often associated with Eastern philosophies and religions. This doesn't mean that everyone living in Asia or India believes in reincarnation, any more than all New Yorkers love hot dogs. But it is safe to say that adherents to the Eastern "way," whether Chinese, French, American, or Italian, will embrace

Fig. 8.2 ©The New Yorker Collection 1989 J.B. Handelsman from cartoonbank.com. All Rights Reserved.

concepts that are foreign to most Westerners. However, some Westerners have become disenchanted with traditional religion and have adopted reincarnation as an alternative spiritual focus.

Steve C. is a thirty-four-year-old high school baseball coach. His degree in physical training and years of experience developing young minds and bodies have led him to beliefs that are in a category of their own.

Humans evolved in a spiritual way by means of Universal Energy. I believe there is life on other planets and it was there prior to life on Earth. We were created separately and connected by Universal Energy—not through evolution. I don't think of a Creator or a God, but Energy.

My beliefs have come from my life experience. I could feel the pain of my students. Then I got into yoga and meditation, and became more in tune with myself. I'm no longer afraid of dying. We are all here for a reason. We chose our lives before coming here. We chose our parents. We are here to work on certain goals. I believe in life after death. The body is just a vehicle for the spirit or soul. I don't know much about karma, but if karma is bad, and we come to our present life to work out the bad, how did karma start in the first place?

We've been around a long time and haven't yet found out what the soul is. I think maybe we're not meant to find out. It's the one thing that we won't be able to prove really exists until we die.

Sophisticated reincarnationists like Steve believe that human souls migrate only into other human bodies. But in popular culture many folks think it's possible for human souls to relocate into the bodies of other animals. A friend whose sister recently died has taken in a stray cat that she claims bears an uncanny resemblance to her sister. The cat, named Molly after the lost sister, reportedly has the same eye color and disposition as her human namesake, and will only sleep in the sister's four-poster bed.

Einstein once remarked that if there is any religion that can cope with the needs of modern science, it's Buddhism—fourth largest of the world's religions. Buddhists deny the existence of a personal creator or an autonomous soul that animates and outlasts the body. The Buddha broke with the Hinduism of India by teaching reincarnation and the role of karma, but not the transmigration of souls. Instead he said that only wisps of the self endure beyond death, never showing themselves but joining the eternal flow of cosmic consciousness that seeks release in the bliss of Nirvana. The Buddha would not have been conned by "Bridey Murphy."

Eastern reincarnationists assume that the soul has an infinite past as well as future. This is a very hard concept for Westerners to grasp. It differs significantly from the traditional Judeo-Christian notion of a soul that began at birth. The soul's ultimate destiny is far more important to Eastern reincarnationists than it is to Western believers, who focus more narrowly on the soul's fate immediately after death. Furthermore, the Western concept of immortality features resurrection of the individual, but not the continuing transmigration of souls into other personalities. Of course there are as many variations on these themes of soul, immortality, and reincarnation as there are believers.

Do past life regressions offer proof of reincarnation? Spontaneous recollections of past lives are rare even in India, where belief in the transmigration of souls is a way of life. Hypnosis is the most successful tool for recalling a past life, but careful investigation consistently traces this phenomenon to *cryptomnesia* or "hidden knowledge."

Our minds continually collect bits of information from real life experience and store them deep within the subconscious. Like a good housekeeper who tries to keep everyday clutter isolated, the mind soon becomes overwhelmed by this information and must clean house. It reorganizes and discards pieces of hidden knowledge through dreaming and hypnosis. Tales of past lives are often the bizarre results of this psychological spring cleaning. The alleged beneficial value of regression therapy may stem from the release of psychological tension when hypnotic suggestion forces unpleasant memories and hidden knowledge to the surface. Tales of past lives don't prove reincarnation—only our suggestible will to believe.

Hypnagogic and hypnopompic hallucinations are other oddities of the human psyche that are commonly accepted as proof of a supernatural realm. These phenomena are experienced by normal people when they first fall asleep or are just waking up. If the transition between sleeping and fully awake is irregular, we can be awake but dreaming. We are unable to move because a temporary state called sleep paralysis prevents us from being able to get up out of bed. During this anomaly of consciousness, we may dream intensely and become frightened because we can't move or flee. Historically people have ascribed these hallucinations as visits from ghosts or demons. Often religious or spiritual needs become involved in the victim's interpretation. These incredibly realistic hallucinations have been implicated in many paranormal phenomena including visits from aliens.

General George S. Patton's lifelong belief in reincarnation was reinforced by his intense experiences of *déja vu*. We have all been swept up in a sudden feeling that we have "been here before" in a place or among people from a dimly remembered past. *Déja vu* is a frequent and powerful mental phenomenon that many believers say is the best evidence for reincarnation.

Extensive scientific study of *déja vu* has revealed that the sensation is caused by a time discrepancy between memory retrieval and recognition, a millisecond short-circuit between left and right-brain processing. The mystery of the déja vu experience often resolves itself once our memories of past events catch up with our senses.

Isaac Newton was born on the same day in 1642 that Galileo died. Reincarnation or coincidence? Perhaps the greatest conceptual impasse to belief in reincarnation is logistics. Where does the soul go in between each mortal appearance?

The period between the soul's abdication of a body at death and its rebirth in a new one has been referred to as the *interregnum*. Duration of this time-out for the soul varies. Many Buddhists and Zen masters feel it's no more than six or seven days, while Plato said it lasts one thousand years. For the soul of "Bridey Murphy," who died in 1864 and was reborn as Virginia Tighe in 1923, the interregnum was only fifty-nine years. Many philosophers have suggested that the souls of very bright and cultivated people (like philosophers) may wait it out as long as ten thousand years. So Newton probably wasn't Galileo after all!

Where does a soul reside while it's "off duty"? There seem to be two possibilities. Plato thought the soul exists as a type of disembodied mind. The alternative to this pure mind, and the one favored by most reincarnationists, is the *astral body*—a non-physical body double that exists in another dimension.

Projections of astral bodies into the visible dimension have been the basis for claims of people appearing in two places at once or body doubles seen as shadows in mirrors. Of course not everyone can see these astral projections, as we learned from the hit movie "Sixth Sense." Astral bodies have never been reliably recorded or objectively studied. They aren't real in any sense of the word. Even the best documented cases of a body double have surrendered to rational explanations once the hype and hope have been separated from the evidence.

Unlike astral projections, out-of-body experiences and the near-death allure of a bright white light are genuine mental phenomena. They can be experimentally conjured up using drugs, hypoxia, and deep hypnosis—the psychologist's crowbar into the subconscious.

Sane, but highly suggestible, people can fall victim to ghostly delusions as easily as the psychologically disturbed. In a famous trial in 1876 the two greatest naturalists of the century, Charles Darwin and Alfred Russel Wallace, took opposing sides in the successful prosecution of Henry Slade, an American psychic who claimed to be in touch with the souls of the dead. Spiritualism had become a popular refuge for many Victorians seeking relief from the cold realities of evolution and physical science. Wallace appeared for Slade's defense, not out of religious belief but from a lifelong interest in spiritualism and a desire to prove the existence of the soul. Darwin gave aid to the prosecution, but declined to appear publicly in court. Slade's fraudulent seances were exposed by the testimony of a young zoologist named Edwin Ray Lankester. Ironically, Lankester would later become Director of the British Museum of Natural History where he was completely taken in by Piltdown—the most notorious fraud in evolutionary biology.

Belief in a world inhabited by spirits and propelled by unseen forces doesn't come out of nowhere. Animistic belief is older than any religion. We assign spiritual forces to animals, trees, rocks, and inanimate objects. We carry them for good luck, thank them for our triumphs, and blame them for our troubles. Our instinctual sense of the spiritual is the result of three million years of inventing answers to the unknown. It is a survival trait that has kept us psychologically strong in a chaotic world.

Our common experience of dreaming that a part of us is awake and moving about reinforces belief in a personality or soul that is separate from the body. Illness, high fever, mental disease like schizophrenia, and drugs like LSD bring forth similar states of dissociation. Our body and soul seem to part company as we plunge into what American philosopher Joseph Campbell called the tides and undertow of the inward sea of our unconscious. Freud said that this "oceanic feeling" is often followed by sudden clarity in understanding and a sense of *knowing* things that had been mysteries. We have now entered the realm of the mystics.

Mystical Moments and You

According to national surveys in the United States and England, one out of three adults say they have had a mystical experience —a moment of religious awakening or spiritual insight.

Before you turn off and tune out to all this mystical talk, ask yourself this question. Have you ever experienced a sense of exhilaration immediately after a great idea hits you—and just before you realize what's wrong with it? At that instant of heightened euphoria, physical objects may recede from view as our abstract thoughts become concrete. You experience a higher reality, a clarity of thought, a visceral understanding of truth. Congratulations! You have just had a mystical moment.

It is difficult for those of us schooled in the Western tradition of experimental inquiry to fathom a path to truth and knowledge that includes mystical insight. American philosopher and psychologist William James recognized the power of mystical experiences and warned,

> *They elicit from us a reaction; and the reaction due to things of thought is notoriously in many cases as strong as that due to sensible presences. It may be even stronger.*

Mystical moments occur during altered states of consciousness in which our senses and logic conspire to reveal unfamiliar dimensions of the mind. *Déja vu* is perhaps the most common mystical experience. But a runner's "high" or a musician's rapture during an intoxicating melody also qualify.

William James claimed that mystical experiences are legitimate topics for scientific inquiry, and described their four key characteristics:

1. ***Ineffable***—They defy description. We might say "you would have to experience it to understand it."
2. ***Insightful***—They reveal significant knowledge unreachable through normal thought processes. We might say we now "know" something important about life, but can't articulate what it is we have learned.
3. ***Transient***—They last only minutes to an hour at most. Memories of their exact nature quickly fade, but are readily recognized if they recur.
4. ***Passive***—Whether we enter them accidentally (*déja vu*) or voluntarily (meditation, chanting, dancing), mystical experiences make us feel powerless in the face of a superior knowledge or unseen force.

Zen Buddhists begin their spiritual quest by learning meditation techniques that enable them to voluntarily enter *satori*—a mystical state that brings an indescribable clarity of thought and a joyous feeling of being at one with a higher reality. Other techniques to induce mystical moments include the drugged trances of shamans, ritual dancing of Islam's whirling dervishes, incessant chanting and drumming of Native Americans, and hypnosis.

Primal religions grew out of oral wisdom traditions that encouraged our ancestors to be open to "knowledge" gained through mystical experience. Immanuel Kant argued that remnants of this primal mystical knowledge survive in our myths, traditional religions, and sense of the sacred. They are given names like God, soul, the design of creation, and life after death. They have no factual basis, but play a practical role in our moral consciousness. They also influence our beliefs about where we came from.

Mystical moments imprint an indelible reality of the unseen within our subconscious that is impervious to facts. Some researchers believe they are products of the same kinds of mental errors that manufacture meaning from random events and enable us to see what we want to believe.

But other psychologists propose that consciousness itself is not a single level of awareness that is either on or off. They argue that awareness operates on a spectrum of levels that vary in intensity from person to person and are culturally determined. Not everyone dreams of flying or has an out of body experience. But most of us will admit to at least one incident when something happened to us that was outside the ordinary realm of conscious awareness. The effect of these events are long lasting and can change peoples' lives.

Arguably the best scientific study on the long term effects of mystical experience was conducted during the sixties by physician and Harvard Divinity School student Walter Pahnke. In 1962 Pahnke assembled twenty Protestant seminary students and randomly assigned them to two groups. One group ingested psilocybin, the active drug in hallucinogenic mushrooms. The other group received a vitamin placebo that made their skin flush. Then both groups attended a Good Friday church service.

Immediately after the service, the group that had received the drug described seeing visions and feeling a sense of spiritual enlightenment—reactions similar to mystical moments. The control group given the vitamin reported no unusual occurrences.

Researchers conducted follow-up interviews with study participants at six months and twenty-five years after the experiment. In both cases those in the drug group reported many more positive changes in attitude and behavior that they directly attributed to the Good Friday sermon than did those in the placebo group. But not all children of the Age of Aquarius were caught up in mystical wonder.

Martha F. is a fifty-five-year-old professional business woman and flower child of the sixties. She retains a style of independent thinking, forged during those early years, that dominates her view of mysticism.

I don't think there is any mysticism. I lived in India for two years and encountered a lot of the mysticism of the East. But there's no more mysticism in the East than there is in the West. It's just different. Living in India was like living on the moon in comparison to what I had experienced in America. It's exciting, exotic, and interesting. Mystical it ain't.

The basis of Hinduism and Buddhism is achievement of Nirvana or nothingness, which comes closest in any religion to what I believe. I don't believe in the soul. It's a nice idea, but if you believe in the soul you believe in the hereafter, and I don't believe in the hereafter. When we're here, we're here. And when we're gone, we're gone.

Today we tune in to our world through telescopes and microscopes—not trances and mystics. We snap pictures of the Universe and explore the heavens beyond any sense of the sacred. But our primal attraction to the reality of the unseen endures, providing solace from the cold reality of science.

Science Searches for the Soul

Brain science pioneer Santiago Ramon y Cajal referred to the pulsating networks of the hundred billion neurons in a human brain as the "beating wings of the mysterious butterflies of the soul." We think of consciousness as palpable but elusive, residing somewhere in the brain, but not wholly of it. What is consciousness and what is it made of?

Consciousness is private and subjective. For this reason alone we may never be able to fully understand it. For most people consciousness means awareness, focused thought, and reasoning. But for others the subjective subtlety of consciousness serves to confirm their belief in a soul. While the soul remains an article of faith, human consciousness has become a hot topic for scientific inquiry.

The accepted view of modern neuroscience is that human intellect and consciousness emerges out of the complex electrical and chemical signaling in our brain. Biologist Francis Crick sums up his view of the omnipotence of this mind-brain dependence in his book, *The Astonishing Hypothesis*, writing,

> *"You," your joys and your sorrows, your memories and your ambitions, your sense of personal identity, and free will, are in fact no more than the behavior of a vast assembly of nerve cells and their associated molecules. As Lewis Carroll's Alice might have phrased it: "You're nothing but a pack of neurons."*

Many scientists agree. They say mystical encounters, near-death events, out of body experiences, and other spiritual phenomena are simply altered states of consciousness that convince us of the reality of the unseen.

Altered states of consciousness can be reproduced and tested in the laboratory. They are characterized by intense feelings of pleasure or a focused awareness that can last for minutes or hours. While in them, we seem to lose track of time. Significant physiological changes take place as we move from one state to another. These are detected as shifts in heart rate, respiration, eye movement, motor activity, voice, and brain waves. For example:

- Brain scans of Buddhist monks in deep meditation show distinct changes in brain chemistry, including lowered activity in areas of the brain responsible for orientation and a sense of place. This could explain the sensation of release into an unfamiliar realm of time and space that the monks interpret as spiritual transcendence.
- Investigations of claims of mystical and paranormal experiences in patients with epilepsy have linked these sensations to bursts of electrical activity in the temporal lobe.
- Altered states of consciousness caused by drugs can be experimentally triggered. For example, the cocaine addict's high can be brought on by administering a placebo. Researchers believe that neurons affected by the

> cocaine form a path in the brain. The addict's consciousness is awakened by the placebo and automatically speeds down this path to reach the familiar state of intense pleasure.

Daydreaming is also an example of a dissociative state of consciousness in which our body is in one place and our awareness in another. Dissociative states can be induced through hypnosis and deep meditation, but they voluntarily emerge in pathologies such as multiple personality disorders.

In *Myths to Live By*, the American philosopher Joseph Campbell points out striking similarities between LSD-induced visions, dissociative disorders of schizophrenics, and the vision quests of ancient shamans in trance states, and asks,

> *What is the difference between a psychotic or LSD experience and a yogic, or a mystical? The plunges are all into the same deep inward sea. But there is an important difference. The difference—to put it sharply—is equivalent to that between a diver who can swim and one who cannot. The mystic, endowed with native talents for this sort of thing, enters the waters and finds he can swim; whereas the schizophrenic, unprepared, unguided, and ungifted, has fallen or has intentionally plunged, and is drowning.*

Campbell sees in these waters the universal archetypes of mythology—the prototypes for the gods, rites, and beliefs found in every human culture. They are products of the mind which enlighten the mystic but devastate the uninitiated.

Carl Jung said mystical states emerge from the collective unconscious in all human beings, explaining, "The basal depth or layer of the psyche is an expression of the instinct system of our species, grounded in the human body, its nervous system, and wonderful brain."

Altered states of consciousness appear as discrete, quantum shifts in physiology, suggesting to some researchers that another dimension is involved beyond our five senses and the hard-wired neural networks of our brain. They believe that consciousness is produced by the same kinds of quantum energies and subatomic vibrations that have been implicated in other functions such as sight and hearing.

Moreover, exciting new research has revealed a distributed brain processing web where input from our senses is integrated with functions such as image formation and memory storage. Sensory and logic data riding this deep brain Internet might simulate the reality of consciousness in much the same way that the digital data of a hologram simulates the reality of three-dimensional objects.

These findings have led researchers like Karl Pribam to take a quantum leap of their own and suggest that in the beginning, there was consciousness—or at least the quantum infrastructure to support it. As we evolved into humans, we were able to tap into this quantum energy of the Universe. Consciousness had found a home in the human brain. If all this sounds a little familiar, it is. Practitioners of Zen Buddhism believe that the solitary "I" is really an emergent property of the mind, an illusion nurtured by a function in our left brain as it taps into the Universal Conscious.

A smattering of psychologists hint that conscious communication may be a two-way street. Our brains may receive mental energies and thoughts, then project them back out into the Universe. There is no evidence for this. But there is also no shortage of believers in this brain broadcast system known as *psi* or paranormal psychic power.

Susan Blackmore, parapsychological investigator and skeptic, has observed scientific experiments of psychic phenomenon and investigated claims of their success for over thirty years. An out-of-body experience launched her on a personal crusade to prove to skeptics and scientists that consciousness has power of its own and can exist beyond death. But now she has thrown in the towel. After years of personal observation and participation in various experiments thought to be totally objective, she has now shown them to be false, flawed, or fraudulent at worst—and the product of self-deception and wishful thinking at best. Blackmore now believes that psi doesn't exist.

So far the most credible studies of paranormal psychic power have been *ganzfeld* tests, first carried out in the mid 1970s by reputable scientists in academic settings. Many of these tests have been shown to have had serious design flaws that jeopardized the reliability of results. In these experiments a "receiver" lies comfortably in a room, listening to white noise or relaxing sounds through headphones while wearing halves of pin-pong balls over their eyes. They see nothing but a uniform white field or *ganzfeld*. The "sender" sits in a remote room, looking at a picture. After a half hour or so the receiver is presented four pictures and asked to choose the one he or she feels was the "target" viewed by the sender.

Some ganzfeld researchers have published claims of large numbers of positive results. Their receivers guessed the right target more often than predicted by chance alone. But other investigators repeated the same experiments and reported no evidence for psi. Questions of sensory leakage have arisen and control measures have been tightened. A few recent successes are tantalizing. But the lack of consistency in results continues to fuel the debate.

What does this have to do with consciousness? Apparently not much. Successful receivers are unable to predict which answers are hits and which misses. If ESP is occurring, the subject is unaware of it. In other words, there appears to be no connection between *psi* and consciousness. Though receivers often lapse into a slightly altered mental state, the involvement of consciousness in psychic power is still unknown.

Unfortunately, any hint of credible evidence for the existence of psi in any of its manifestations, whether remote viewing, bending spoons, or fortune telling, will be taken by believers as proof that consciousness exists independent of space and time, and has powers that reach beyond the brain. And as Blackmore points out,

> *I suspect that it is a desire for this "power of consciousness" that fuels much enthusiasm for the paranormal. Parapsychologists have often been accused of wanting to prove the existence of the soul, and convincingly denied it. I suggest instead that parapsychologists want to prove the power of consciousness... They want*

to find that consciousness can do things all by itself, without dependence on a complicated, physical, and highly evolved brain.

That kind of dangerous retreat to Descartes' dualism and complete separation of mind and body would put us on the slippery slope of irrational science, tumbling headlong toward justification for belief in the unseen.

Despite exciting new discoveries that have shown there is more to the mind than the matter of the brain, most neuroscientists insist that the human consciousness or soul is a manifestation of deep brain networks. But theologians caution that reducing the soul to brain synapses and quantum signaling only scratches the surface of spirituality. As one believer rationalized it, "If we recognize the brain does all the things that we (traditionally) attributed to the soul, then God must have some way of interacting with human brains."

Michael Arbib, an expert on brain theory at the University of Southern California, counters, "We cannot approach theology without some sense of the intricacy of the human brain. A lot of what people hold as articles of faith are eroded by neuroscience. The mind has properties—self-consciousness, wonder, emotion and reason—that make it seem more than material. Yet I argue that all of this can be explained eventually by the physical properties of the brain. In twenty years, we will understand what happens in the brain when people have religious experiences."

Some scientists have already found a personal connection between science and the sacred in their explorations of the most fundamental forces in the Universe.

Where East Meets West

What do scientists believe about God and immortality? In a classic survey conducted in 1916, psychologist of religion James Leuba found that 40% of scientists believe in a personal God. In 1997, Edward Larson and Larry Witham repeated Leuba's study and got similar findings. Many scientists felt that this measure of personal belief among scientists was too high because Larson and Witham's survey could not distinguish true believers from those who attended church for the sake of families or out of a sense of tradition.[2]

In a follow-up study of eminent scientists who were members of the National Academy of Sciences, belief fell to under 30%. When agnostics and other doubters were factored in, the extent of true belief virtually disappeared. Did these numbers reflect personal belief or the peer effect urging distinguished men and women of science to follow the party line?

These studies reflect only the beliefs of American scientists. Surveys of religious belief in the United Kingdom, Europe, and other developed countries consistently report much lower levels of personal religious belief among both scientists and the general public.

Belief in immortality was a very different matter. In 1916, 51% of American scientists believed in immortality, and only 20% did not. By 1997 belief had declined to 38% and disbelief had risen to 47%. During this same period, the number of agnostics was cut in half. Advances in scientific knowledge have apparently calmed the sea of doubt for many scientists, but more than one out of three still hold out hope for eternal life.

Some people think science should be able to explain everything —and so do many scientists. In his impassioned 1976 best-seller, *The Tao of Physics*, theoretical physicist Fritjof Capra kicked-off a trend among cosmologists and physicists to find a sense of the spiritual at the frontiers of science. In his book, Fritjof drew parallels between the abstract language of modern physics and the prose of Eastern mystical texts such as the stories of the Hindu Vedas and the philosophy of the *Tao Te Ching*. He proposed that these resemblances were not coincidental but describe the same truths, including the unity of the atomic structure of the Universe and Einstein's theory of relativity. Fritjof's book hit a chord with the public, but reaction from the scientific community was more cautious. Most scientists dismiss Fritjof's analogies as coincidence, not revealed gospel.

In 1973 cosmologist Brandon Carter introduced the "anthropic principle," a theory that claims the laws and physical constants governing the Universe are so precise that if they were to vary by the smallest fraction we would not be here. But life is possible and we are here. Carter, Frank Tipler, and other proponents of this argument see it as proof that an intentional Creator made the world just for us.

Stephen Hawking, possibly the most brilliant mind on the planet, expressed a different sense of awe in his best-seller, *A Brief History of Time*. He anticipated the discovery of a unified theory that will explain all the forces in the Universe, writing,

> *If we do discover a complete theory, it should in time be understandable in broad principle by everyone, not just a few scientists. Then we shall all, philosophers, scientists, and just ordinary people, be able to take part in the discussion of the question of why it is that we and the Universe exist. If we find the answer to that, then it would be the ultimate triumph of human reason—for then we would know the mind of God.*

Many scientists feel a sense of wonder when they contemplate the natural world. But the scientist-skeptic uses empirical reasoning—not mystical insight—to find the answers. Lest anyone mistake the intent of Hawking's words, his answer when directly questioned about his belief in a personal Creator is a firm "No."

But in his book *The Mind of God*, physicist Paul Davies spins Hawking's words to make a very different point. Davies examines the great questions that have preoccupied philosophers and scientists throughout history, then reviews the evidence to conclude,

> *We are barred from ultimate knowledge, from ultimate explanation, by the very rules of reasoning that prompt us to seek such an explanation in the first place . . . I cannot believe that our existence in this Universe is a mere quirk of fate, an accident of history, an incidental blip in the great cosmic drama. Through conscious beings the Universe has generated self-awareness. This can be no trivial detail, no minor byproduct of mindless, purposeless forces. We are truly meant to be here.*

Davies uses the term God for this deeper level of explanation beyond science—which he admits is a personal, emotional choice and not proven. But a degree in quantum physics isn't a requirement for holding complex beliefs about matter, energy, and the beginning of life.

Jim R., a professional gardener in his sixties, rejected his Catholic faith during college and married an atheist. He now describes himself as a liberal Christian. Over the years Jim's beliefs have matured into a blend of Western and Eastern themes.

I believe we came from a universal force that some call the River of Life. I call it God. In the beginning we were all part of God, then became separated. The purpose of life is to go back and be one with God.

I don't know when the soul enters the body, so I am undecided about the abortion issue. Some experts say the soul and near death experiences are just chemical reactions in our brain. I believe in evolution, and that we are still evolving both physically and spiritually. We're not the complete package yet. As for understanding of reincarnation, humans are at the level of kindergarten.

I hope I live long enough to see a different form of life from another planet, because reincarnation doesn't mean that souls return only to the Earth.

By the way, my Survey participants aren't a very spiritual group. Over half confess to having little or no active involvement in spiritually-oriented gatherings and flatly deny reincarnation. One out of four haven't made up their minds about it.

But the Creationists have no doubts. For them, life began as a miracle, and we are not the mundane product of some ill-behaved process of evolution that operates "in fits and starts."

NOTES

1. George Gallup, Jr., "Religion in America," *The Public Perspective*, Vol.4, No. 3, The Roper Center, University of Connecticut, 1995, pp. 1–8.

2. James H. Leuba, *The Belief in God and Immortality*, Open Court Publishing Co., Chicago, 1921; Edward J. Larson and Larry Witham, "Scientists are Still Keeping the Faith," *Nature*, Vol. 386, April 3, 1997, pp. 435--436; George Bishop, "The Religious Worldview and American Beliefs About Human Origins," *The Public Perspective*, Vol.9, No. 5, The Roper Center, University of Connecticut, 1998, pp. 39–44.

REFERENCES

Blackmore, Susan. 2001. "What Can the Paranormal Teach Us About Consciousness?" *Skeptical Inquirer* 25(2):22–27.

Capra, Fritjof. c1975 (reprint 1984). *The Tao of Physics: An Exploration of the Parallels Between Modern Physics and Eastern Mysticism*. New York: Bantam.

Crick, Francis. 1995. *The Astonishing Hypothesis: The Scientific Search for the Soul*. New York: Touchstone.

Edwards, Paul. 1996. *Reincarnation: A Critical Examination*. Amherst: Prometheus.

Gallup, George, Jr. 1995. "Religion in America." *The Public Perspective*. 4:1–8. University of Connecticut: The Roper Center.

Hawking, Stephen W. 1988. *A Brief History of Time*. New York: Bantam.

James, William. 1958. *The Varieties of Religious Experience*. New York: New American Library.

Larson, Edward J. and Larry Witham. 1997. "Scientists are Still Keeping the Faith." *Nature* 386:435–436.

Leuba, James H. 1921. *The Belief in God and Immortality*. Chicago: Open Court Publishing.

Shanor, Karen N. 1999. *The Emerging Mind*. Los Angeles: Renaissance.

WEBSITE

Scientific Approaches to Consciousness
(http://www.nimh.nih.gov/events/consciousness.cfm)
Excellent summary and video of the May, 2000 National Institute of Mental Health symposium on the search for the source of human consciousness. Top scientists and thinkers clarify the many aspects of the mind versus body question. Mind boggling stuff.

SUGGESTED READING

Karen N. Shanor, *The Emerging Mind*, Renaissance Books, Los Angeles, 1999. Current thinking about the meaning of consciousness by a variety of experts. Intriguing information accessible to the layperson. Based on the Smithsonian Institution Lecture Series and edited by a psychologist and popular media expert on matters of the mind.

Paul Edwards, *Reincarnation: A Critical Examination*, Prometheus, Amherst, NY, 1996.
A comprehensive, well-documented discussion of reincarnation by a skeptic. Includes evaluation of evidence, conceptual and scientific objections, and examples of other mind-brain phenomena.

Chapter 9

Evolution in Fits and Starts

Science is proof without certainty; faith is certainty without proof.
Anonymous

Two little girls were on a class field-trip to the American Museum of Natural History in New York. They ran excitedly into a room filled with life-size models of early human ancestors.

"Look, monkeys!" yelled the first little girl.

"They're not monkeys," corrected the other girl. "They're humans. They just look like monkeys because humans *evolved* from monkeys."

"Really?" said the first little girl, obviously amazed.

"Yes, I think so," replied her friend.

Then their teacher rushed onto the scene with the rest of the class in tow. The children stared wide-eyed as she explained,

"These models represent one idea about where humans came from. Some people think we evolved from monkeys like these. But there is another idea—that humans and all animals were created by God."

Who do you think the first little girl believed—her friend or her teacher?

How did you react to this true story from the Big Apple? Were you delighted or dismayed with the teacher's account of human origins? It might surprise you to know that every day in public schools across America a similar version of human origins is being presented in biology and science classrooms. Today there is virtually no debate over teaching that the Earth revolves around the sun. But the teaching of evolution remains an emotionally charged issue.

Alicia T. is a sixteen-year-old high school junior whose biblical fundamentalism has found an ally in her high school science teacher.

I think we need to teach creation as a theory, because it can't be disproved. I had a biology teacher in tenth grade who was a Christian. He taught evolution like it was in the textbook, but said it's only one theory and nobody knows for sure.

Then he said there's another theory that God created people. He gave it pretty objectively, nothing set in stone. It's wasn't part of the text. He just put it in. But he said if you're going to tell the story of human origins you should bring in both sides.

That's not legal, you say? Creationism isn't science, it's religion and you don't want your children taught religion in place of science? In fact, a 1987 Supreme Court decision overturned Louisiana's "Creationism Act" that required the teaching of "creation science" in public schools whenever the theory of evolution was taught. But the battleground simply shifted from the high bench to the high school classroom where creationists now cloak their fight in more subtle pleadings.

From Friendly, Nevada to Marietta, Georgia, creationism has emerged as a contentious issue for local school boards. Religiously conservative board members are an easier sell than Supreme Court justices. Often more open-minded members can also be persuaded to vote for "fairness" in balancing the airtime given to what they believe are two equally well-documented theories of human origins. Instead, they are breaking the law of the land by introducing religion into the classroom. These aren't struggles in a righteous crusade, but skirmishes in a more insidious campaign to control the information and beliefs of society. Who are the creationists and what is their agenda?

God said it, I believe it, and that's that

Surveys confirm that almost half of all adult Americans identify with the statement: "God created humans pretty much in our present form within the last ten thousand years." These are the Creationists who believe that human beings are the direct creation of a supernatural Creator, Intelligent Designer, or God.

National polls conducted during the last twenty years have told us a lot about the social, political and demographic profile of Americans who hold a creationist worldview. Most are older, live in the South, and consider religion "very important" in their lives. They tend to belong to fundamentalist Protestant denominations, such as Baptist, that describe themselves as "evangelical." Creationism doesn't always correspond to religious fundamentalism, but most creationists share an unusually strong respect for tradition and authority. Interestingly, more African-Americans and women are likely to be creationists than whites and men.

America is often thought of as the most technologically advanced nation on Earth. But over the past fifty years, polls have also shown that it is one of the most religious. Today 96% of Americans say they believe in a Universal Spirit or God—a larger percentage than in any major European country except Ireland and Northern Ireland.

Worldwide surveys also report that Americans are among the *most likely* to hold a creationist worldview. We are three times more likely than the Norwegians, and almost five times more likely than the British to believe that "The Bible is the actual word of God and it is to be taken literally, word for word." In fact creationism hardly exists outside America, except for pockets in

Australia, New Zealand, and Russia. A 1993 cross-cultural science survey of twenty-one nationalities found that adult Americans are the *least likely* to agree that "human beings developed from earlier forms of animals."[1]

Why has creationism persisted in America?

The Scopes' trial in 1925 was an aberration to many Americans—a sideshow among a minority of evangelicals in the South where biblical literalism remained entrenched. Through the first half of the last century, organized creationism rarely rose above the level of a few club-like societies that tried to mold scientific facts to fit scripture.

Since the 1980s the growing influence of religiously and politically conservative groups known as the New Christian Right has rekindled a creationist worldview that harks back to 19th century America. Despite the U.S. Constitution's Establishment Clause separating church and state, we rally around prayer in schools and hang the Ten Commandments in our courtrooms. In this cultural and political climate any public acknowledgement of non-religious beliefs—including belief in evolution—can lead to social ostracism. A spiral of silence builds and muffles the voices of reason.

The higher the education level, the more likely a person is to accept evolution. However, despite a rise in the overall educational level of Americans, the extent of creationist belief has remained surprisingly constant since 1981 when Gallup first began regular nationwide polling on beliefs about human origins. In some surveys as many as one out of four people holding graduate degrees are creationists.[2] As with any social movement, creationism cuts across social class and pops up in unlikely places.

Tim N. is a successful international businessman who was, by his own admission, a wild liberal during his college years. But a mid-life crisis and personal quest turned him around 180 degrees toward a different view of reality.

I have had an inner awareness of a supernatural being and reality that transcends what we can see and touch. That awareness found a center in the Bible of Christianity, after I checked out practically everything else available. I needed a Savior. He came first, "creationism" came later.

My reading in creation science is not extensive, but what I have read seems quite plausible and has as much evidence going for it as evolution.

What do creationists believe?

Creationists not only deny that humans evolved, but find evidence for human origins and the beginning of life in a literal reading of the two-thousand-year-old story of Creation in the book of Genesis. And that's the first problem.

If we're going to formally accept and teach a creation myth, why not the one about the creation of the Hero Twins—the first humans and mythical founders of the Navajo Nation? After all, when you need a creation myth, why isn't one just as good as another? Freud placed the myths of religion on a psychological par with dreams—manifestations of unconscious fears and neurotic compulsions. Our neuroses that shame us in private are accepted in public when projected as gods.

In contrast to Freud's negative take on religion, psychologist Carl Jung valued the ability of myth and religion to take us away from the externally-focused consciousness of our daily grind and pull us back into the strength of our human psyche. But Jung also warns,

> *There is a danger of being drawn by one's dreams and inherited myths away from the world of modern consciousness, fixed in patterns of archaic feeling and thought inappropriate to contemporary life.*

Jung suggests that the only way a culture can escape this danger is to constantly update what we know as "truth" by allowing an interplay between our spiritual beliefs and the modern insights of conscious reason, objective science, and new evidence. As we do this there will come a point where reason must part company with faith.

But this is not the case with creationism, a worldview that differs primarily in its assumptions about knowing. Creationist bumper stickers proclaim, "If God said it, I believe it, and that's that." Creationists "know" that the Bible is historical fact based on divine inspiration. Their unquestioning belief in the truth of the Creation of all life including Adam and Eve, the first humans, becomes the standard by which all other theories are judged—especially evolution.

Susan Losh has revealed this upside-down logic in her study of forty Florida congregations, many of them fundamentalist denominations. She found that these believers interpret the world around them by quoting the appropriate chapter and verse from the Bible, not by forming testable theories. For example, they might believe that humans and dinosaurs were created in the same 24-hour day, though a few hours apart and in different places on Earth. As evidence they cite the passage where God says to Job, "Look now at the behemoth which I made along with you." (Book of Job 40:15)

Above all creationism is a belief about morality. Sociologists describe Creationists as cultural traditionalists because they look to formulas from the past, often found in a literal reading of the Bible, to solve social problems and answer moral questions of today. Immersion in the psychology of group dynamics binds these believers and generates a distinct morality. Any discussion of human origins is very controversial because it threatens the creationists' core values.

A recent survey of first-year medical students in Melbourne, Australia found that nearly 20% believe all forms of life arose from ancestors who rode out the great Flood on Noah's ark. Why has creationism persisted in our highly technological society? Studies have shown that students who enter college with creationist beliefs will leave holding essentially the same worldview. Their beliefs are impervious to education in science and logic, or exposure to overwhelming evidence to the contrary. It's not ignorance at work here but the effect of deep psychological and social programming. Creationism is so tightly woven into their self-image, core values, and view of reality that it acts as a shield against contradictory evidence. From time to time talented scientists who are creationists have attempted to slip their beliefs into their work. Most openly ally themselves with colleges and research institutions where they find support for their beliefs.

All societies have at some point attributed human existence to an abrupt act of creative intent by a divine force that transcended the laws of nature. The myths and legends surrounding these events have become canonized as religious scripture. Early attempts to reconcile biblical scripture with science quickly revealed formidable obstacles. But ascendancy of Newton and the modern scientific worldview did not cause a profound religious crisis in science. In Darwin's day many scientists accepted evolution while still maintaining a belief in divine intervention. The term "creationist" didn't exist then, and most naturalists would not have been part of the modern creationist movement.

But on the heels of the *Origin of Species* it became more difficult for biblical literalists to reconcile a young Earth wiped clean by Noah's flood with scientific evidence for a very old Earth filled with layers of extinct forms of life. In 1860 an obscure writer named Isabella Duncan claimed to have found the answer. She said there had been *two* creations. In her book *Pre-Adamite Man: The Story of Our Old Planet and Its Inhabitants* (1860), Duncan embraced the geological and fossil evidence, conceding that a biblical day could be a geological age. But recognizing that no human remains had been found among the oldest fossils, she suggested that God had created pre-Adamites during the first creation, then destroyed all trace of them (perhaps with Noah's flood) before carrying out the second creation. Duncan was a biblical literalist, but respected scientific knowledge and tried to reconcile it with her beliefs.

In the 20th century similar attempts to put a scientific spin on religion gave birth to a uniquely American movement known as "creation science"—biblical creationism repackaged as science.

The Evolution of Creationism

During the early 1900s George McCready Price, a devout creationist, nearly succumbed to the siren song of science. But prayer and exhaustive study of the teachings of Ellen G. White delivered him from the jaws of evolution. White was a founder of the fundamentalist Seventh-Day Adventist sect. The importance of the Sabbath Day to Adventist beliefs had inspired her to reject any interpretation of Genesis that equated a day of creation to a geological period. White challenged believers that a true reading of Genesis, uncluttered by the principles of geology, would reveal the truth of Creation. George Price did exactly that and soon found what he was looking for.

Price revoked the slow, steady natural processes that had sculpted the planet's surface, and reconstructed the science of geology based on biblical evidence to claim:

- The Deluge had been caused by a sudden shift of the Earth's axis that released massive amounts of subterranean water.
- This worldwide disaster killed smaller animals first. Their subsequent burial and position in the fossil sequence was determined by how high they floated. Vertebrate fishes floated to the top. Larger animals and humans fled to the hills—where they died but were buried later by a miraculous cosmic storm.
- Great mountain ranges like the Himalayas formed from mounds of debris deposited by the Flood, then folded up to their present height by the pressure of subsiding flood waters.
- Before Flood debris hardened into the layers of silt and rock we see today, natural forces quickly carved out formations such as the Grand Canyon.

This was Flood Geology at it's best.

Price became a one-man band. He eagerly goaded the scientific establishment into acrimonious debates and preached his illogical geology to anyone who would listen. He published articles and mailed pamphlets to prominent scholars. Later in life Price recanted some of his earlier theories, but the basic thrust of his arguments remained.

For the next two decades leaders in the creationist movement clashed over the details of flood geology. As in any social movement, these true believers were an eclectic fraternity. Many were eager to meet science head-on with Price's Deluge theories, no matter how ludicrous. Others thought that reconciliation with the facts of science was essential. Most simply ignored the squabbles and held onto their faith. By the 1930s creationism in America had evolved from a one-man crusade to small armies like the Religion and Science Association. These groups sought to present a united front against evolution, but were fractious and short-lived.

In 1938 a then aging Price formed the Deluge Geology Society. In order to prevent the internal dissension and self-destruction suffered by earlier groups, Price limited membership to those who swore an oath stating:

- The week of Creation lasted no more than six literal days
- The Deluge should be studied as the cause of all major geological changes since the Creation.

As the ranks of the Deluge Geology Society swelled to six-hundred, its members undertook serious field research. Their initial foray was a failed attempt to locate Noah's ark.

At the turn of the century dinosaur tracks had been found along the Paluxy River near Glen Rose, Texas. Though the giant footprints were hard to discern, some investigators speculated they also spied human prints alongside those of the dinosaurs. Several newspapers and magazines picked up the story. A *Scientific American* article predicted that "all the geologists will resign their jobs and take up truck driving if such prints were found to be human." The Paluxy human prints were soon discredited by everyone except the Deluge Geology Society. Its newly formed Footprint Research Committee hastily convened to give its first report and keep hope alive.

Meanwhile as science and technology advanced, creationists could no longer ignore the evidence of geology and the fossil sequence. Attempts to reconcile science with a worldwide flood became less viable. Some creationists began to accept the possibility of small, limited changes in species. Rebellion among the ranks drove more holes in flood geology.

But a far more dangerous enemy of creationism was *radioactive dating*—a new technique that accurately measured the age of the Earth. The frightened president of the Deluge Geology Society complained that radioactive dating threatened to "place one billion six-hundred million years between verses 1 and 2 of Genesis." By 1947 internal dissension over a young Earth versus an old one spelled an end to the Deluge Geology Society. Radioactive time had claimed its first victim.

No one has done more to promote the creationist creed in late 20th century America than Henry M. Morris—godfather of "creation science." The Texas-bred Southern Baptist was a talented engineer from Rice University with a brilliant future ahead of him. While earning his doctorate at the University of Minnesota, Morris reconnected with creationism by joining the American Scientific Affiliation. This group billed itself as committed to creating and distributing accurate information on the connection between science and religion. All its members swore an oath on the Bible that read,

> *Since God is the Author of this Book, as well as the Creator and Sustainer of the physical world about us, I cannot conceive of discrepancies between statements in the Bible and the real facts of science.*

Morris soon became an outspoken creationist and never let up throughout a promising career. While head of one of the nation's largest civil engineering programs at Virginia Polytechnic Institute, Morris began a joint collaboration with John Whitcomb of the American Scientific Affiliation on their creationist blockbuster, *The Genesis Flood* (1961).

Whitcomb's chapters clarified biblical evidence and interpreted details such as:

- *Capacity of Noah's ark* (equal to eight freight trains with sixty-five stock cars each)
- *Source of Deluge water* (an antideluvian canopy of water surrounding the Earth—plus the biblical "fountains of the deep")
- *Distribution of animals* (after the flood they raced across ancient land bridges)

Then Morris highlighted the problems posed by Whitcomb's analyses. He warned Bible-believing Christians they would have to reject either the scriptures or the testimony of trained geologists. Morris recommended the latter.

Publication of *The Genesis Flood* marked the renaissance of creationism and George Price's flood geology. The book was not about the origin of species because as it stated:

> *The geologic record may provide much valuable information subsequent to the finished Creation . . . but it can give no information as to the processes or sequences employed by God during the Creation, since God has plainly said that these processes no longer exist.*

Much of the book's initial reputation was built on Morris' theory that dinosaurs and humans had lived side by side—a specious claim finally removed in the third printing. *The Genesis Flood* sold over 200,000 copies but was virtually unknown outside the world of fundamentalist religion.

Morris' instant popularity prompted a hectic speaking and writing schedule along with his teaching duties. By the late 1960s, however, his rabid creationist views, charges of promoting pseudo-science, and administrative clashes led to an enforced sabbatical from Virginia Polytechnic. But Morris had already decided to go West to southern California where in 1970 he established the Institute for Creation Research (ICR) in San Diego.

The ICR conducted little research but turned out lots of newsletters, books, and pamphlets. In 1981 Morris estimated that over a million copies of books by ICR authors were in circulation, including his own *Scientific Creationism* (1974). The ICR became best known for its lively campus debates featuring prominent evolutionists. No one could deliver the creationist message better than ICR Vice-President, Duane Gish.

Gish was a former research biochemist for Upjohn who became Morris' mouthpiece. He was an aggressive public debater whose verbal thrusts and parries handily dismissed scientific concepts such as punctuated equilibrium as "evolution in fits and starts." Gish borrowed indiscriminately from conflicting concepts to make the case for creationism in his best-seller *Evolution? The Fossils Say No!* (1974). He delighted in hawking creation science's Top Ten:

1. Life, the Universe, and all energy were created suddenly out of nothing.
2. The Earth is only six thousand years old.
3. All geological strata worldwide were laid down in a single cataclysmic event—the Great Flood.
4. Originally created kinds of life can change only within fixed limits.
5. Mutation and natural selection are inadequate to cause the development of all living forms from a single organism.
6. There are no "acceptable" transitional fossils connecting one species to the next.
7. Humans and apes have separate ancestry.
8. Humans didn't evolve.
9. Similarities in the chemistry of all life on Earth isn't evidence for evolution. It's to be expected because all organisms inhabit the same world and eat the same things.

10. The increasing complexity of life required by evolution violates the Second Law of Thermodynamics that says all matter moves from order (more energy) to chaos (less energy), not the other way around.

By the way, the evolutionary scientist's answer to that last point is that evolution doesn't violate this or any law of thermodynamics because the Earth is an open system with the Sun providing a continuous source of energy.

Most scientists are shocked and stymied by the creationists' pseudoscience (dinosaur and human tracks together), outright denial of scientific evidence ("you can't show me a single transitional fossil"), and twisted logic ("there's an amazing correlation between the acts of Creation and the facts of nature that disprove evolution"). Many scientists simply dismiss them as crazy and unworthy of response. Others take the bait, but are ill-prepared to explain complex science in stinging sound bites. Their long-winded answers lose the audience to entertaining populists like Gish.

Creation science has a decided advantage in any debate. It's dangerous to dismiss the power of belief in the reality of the unseen and "knowledge" based on certainty without proof.

> Bettie J. is a well-educated, award-winning third-grade teacher who has taught for nearly thirty-five years. She is interested in scientific explanations for the creation of the world and the first humans, Adam and Eve.
>
> *I had a friend who became an ardent apostle for creation science. I listened to him and some of the creationist publicists like Gish and was satisfied with their position. My friend claimed to have had a revelation, a vision that convinced him that the world was created with "apparent age." I have adopted that reality. Things can look a lot older than they actually are, since age is a characteristic of things just like mass, density, chemical composition, or the color of eyes.*
>
> *This is what has fooled the skeptics. They need enormous spans of time for evolution to work. And the apparent age of fossils and minerals leads them to believe there was enough time.*

National surveys during the past twenty years have shown that less than one person out of five holds the fundamentalist creationist beliefs typical of the 19th century.[3] And not all of these believers would claim that divine creation took place in six 24-hour days. Some modern day creationists, disenchanted with the creation science of Morris and Gish, have found a new prophet in astronomer Hugh Ross.

Ross is a new brand of creation scientist. In his book, *The Genesis Question: Scientific Advances and the Accuracy of Genesis*, Ross doesn't deny evolution, geology, or the Big Bang, but freely manipulates both scripture and science to make them agree. For example,

- Ross claims without evidence that life probably arose on Earth at least fifty times between 3.5 and 3.9 billion years ago, therefore it's easy to create life.

  Question: Why have we not been able to make life or DNA out of raw chemicals in a test tube?
  Ross' Answer: Only God can make a tree—or DNA. Life was intentional and Earth was fine-tuned to support it.

- Discovery of disks surrounding several stars leads Ross to conclude that the early Earth had an opaque atmosphere, much thicker than Venus has today.

  Question: Why does a thick atmosphere no longer exist around the Earth?
  Ross' Answer: Because of the "Moon miracle." Over four billion years ago the Moon crashed into the Earth, clearing the debris from our atmosphere.
  Ross' Proof: "And God said, let there be light." (Genesis 1:3)

Ross' readers may be gullible enough to believe this is science, but Ross knows better. Today creation's hi-tech soldiers are wielding a stealth weapon. Their easily impugned notion of "creation science" has given way to the more palatable and slippery concept of "Intelligent Design"—creationism without the G-word.

In his best-seller, *Darwin's Black Box*, biologist and creationist Michael Behe insists that evolution and natural selection can't account for what he calls *irreducible complexity*—a level of organization and precision in chemical systems that can't be explained by current science. Behe infers there must be an Intelligent Designer, a variation on the argument from design adopted by many 18th century naturalist-theologians like Reverend William Paley. They claimed that complex organs, such as the human eye, beg for the existence of a Designer just as a watch needs a watchmaker.

Behe's demand for an Intelligent Designer is a transparent plea for "intelligence divine," and illustrates an element of false logic that all practicing skeptics will recognize. Absence of evidence (in this case a complete explanation for all complexity in nature) is not evidence for the absence of evolution. Nor is it evidence for the alternative (creation).

Unexplainable doesn't mean inexplicable. Behe ignores the reality of modern science that brims with new concepts, tools, and techniques to solve the unsolvable. Complexity theory is in its infancy, but already our knowledge of self-organizing chemical systems and the power of developmental genes is unraveling mysteries that have confounded biologists for two centuries.

Behe and other creationists who advocate Intelligent Design, including Phillip Johnson (*Darwin on Trial* and *Defeating Darwinism by Opening Minds*) and William Dembski (*The Design Inference*), are careful to steer clear of supernatural powers or a personal God in their writings. Most Intelligent Design proponents acknowledge that the design of each species wasn't perfect, and accept evolutionary change within limits. Their purpose becomes clearer when we come face to face with the politics of creationism.

No one has grasped the political and social agenda of the creationist movement in American culture as keenly as Raymond Eve and Francis Harrold, who write,

> *The creation/evolution debate is not primarily one over the scientific status of various accounts of origins. In the scientific research community, these issues were settled long ago. Instead, creationism provides cultural traditionalists with a cause that can bring coherence to their opposition to a wide range of sociopolitical trends such as abortion rights, gay rights, drug use, pornography, equal rights for women, and so on.*
>
> *For these persons belief in evolution becomes the first step down the slippery slope to secular humanism, with its emphasis on flexible ethics derived out of social interaction. Therefore, creationists argue that acceptance of evolution will lead to all the social problems and moral vacuousness seen in contemporary society. From such a perspective, even devout fellow Christians, if they accept evolution, are nearly as bad as secular humanists.*[4]

The goal of the creationist movement is restoration of 19th century religious and cultural traditionalism to 21st century America. Their strategy is to control the hearts and minds of our children. After all, when you've got them as children, you've got them for life.

The Politics of God

Over fifty pieces of legislation have been used unsuccessfully to interfere with the teaching of evolution. In a surreal scenario following atheist Madalyn Murray's successful Supreme Court fight to keep religion out of her son's classroom, California creationists were able to persuade the State Board of Education to revise its 1970 science standards to read,

> *While the Bible and other philosophical treatises also mention creation, science has independently postulated the various theories of creation. Therefore, creation in scientific terms is not a religious or philosophical belief.*

From that moment on it was Katie-bar-the-door for the creationist movement, which quickly delivered a one-two punch. Publication of its high school text titled *Biology: A Search for Order in Complexity* (1970) was followed by a revised version that championed two legitimate theories for the question of human origins—the "evolution model" and the "creation model." Included with the textbook was a teaching aid called *Scientific Creationism*.

During the ensuing California textbook controversy, evolutionists won a Pyrrhic victory by keeping scientific creation texts out of public school classrooms. The uproar and muddled media accounts during the skirmish reduced evolution to a speculative theory and enhanced the image of creationism as science. Another resurgence of creationism during the early 1980s led to a counterbarrage of books from evolutionists defending Darwin's theory and citing the enormous body of evidence that supports it.

During the next decade grass-roots creationism grew into an organized political force. Creationists campaigned to establish "equal time" laws at the state level and voted for conservative majorities on local school boards. In the 1992 elections, a creationist majority controlled twenty-two hundred out of sixteen thousand school boards—nearly 14%.

Today creationism thrives in many local school districts. A blind eye is turned toward well-meaning teachers who feel compelled to insert creationist concepts into classroom discussions about human origins. Numerous studies in 2000 and 2001 confirmed that, despite strong science education standards, many biology teachers do not teach evolution, or do so reluctantly because they themselves are creationists. Others who understand and accept evolution are reluctant to teach it because it is controversial. They don't relish the pressure and criticism they'll receive from creationist parents and school board members.

In states such as Texas and Georgia, where large school districts and conservative school boards wield the power of the wallet, textbook publishers have been more than willing to customize their offerings by deleting chapters on evolution. Politically conservative legislatures are devising novel but legal means around any roadblock. For example, Alabama requires that the following warning be pasted into textbooks:

> "This book may discuss evolution, a controversial theory some scientists give as a scientific explanation for the origin of living things, such as plants, animals, and humans . . . No one was present when life first appeared on Earth. Therefore, any statement about life's origins should be considered as theory, not fact."

The "controversial" theory of evolution is, in fact, one of the most widely accepted, tested, proven, and functional concepts in science. The benefits of modern medicine and advances in agriculture that we all enjoy are based on the biological principles described by evolution.

The Intelligent Design argument has been a godsend for the creationism movement in California and other states where balanced treatment in classrooms is allowed. Its rational presentation of creationist beliefs in scientific language has spawned numerous high school biology texts. Many are written by credentialed scientists and promoted as non-religious. For example, the textbook *Of Pandas and People* describes its central theme of Intelligent Design using this analogy:

> *Darwinian evolution locates the origin of new organisms in material causes, the accumulation of individual traits. That is akin to saying the origin of a palace is in the bits of marble added to the tool shed. Intelligent design, by contrast, locates the origin of new organisms in an immaterial cause: in a blueprint, a plan, a pattern, devised by an intelligent agent.*

A front-page article in *The Wall Street Journal* titled, "Darwinian Struggle: Instead of Evolution, a Textbook Proposes 'Intelligent Design'" referred to the book as "either an unflinching scientific look at flaws in evolutionary theory or the advance wedge of a new effort to return old-time religion to U.S. schools." What do you think?

In late 1999 the conservative Kansas State Board of Education voted six to four to remove virtually any mention of evolution from the standards used to build the state's science curriculum. Now on a roll, the creationist-influenced majority also voted out references to the great age of the Earth and the Big Bang. The vote reverberated throughout the nation, embarrassing Kansas teachers and politicians alike. In the 2000 election, moderate candidates backed by an energized pro-science movement successfully replaced creationist boardmembers. In February 2001 the new Board voted seven to three to revise the science standards and base them on scientific theories accepted by the majority of scientists around the world. Evolution had returned to Kansas.

A recent poll confirms that 83% of Americans support the teaching of evolution in public schools. But that same survey also found that 79% thought creationism had a place in the public school curriculum—the majority saying only as a belief, not a competing scientific theory.[5]

Creationism's popular and political clout relies on its subliminal message that Darwinism is next to godlessness. Creationists claim that evolution is the religion of atheists and the root of a multitude of social evils from communism and crime to inflation and pornography. They argue that these sins of science should not be allowed to gain a foothold in our schoolrooms—not to mention bedrooms, boardrooms, and courtrooms.

Evolution is science, not ethics. Belief in evolution in and of itself could never demolish or discredit a complex system of ethics. As philosopher of science Philip Kitcher expresses it in his book *Abusing Science*,

> *If the issues were not so serious, laughter would be the appropriate response to the Creationists' quick indictment of evolution: We learn of our kinship with other animals, so we turn savage and promiscuous, tear down our social institutions, and abandon ordinary attitudes to personal relations.*
>
> *Still it does seem possible to doubt that the Happy Hooker keeps a copy of "The Origin of Species" tucked under her pillow or that the average mugger draws inspiration from a careful reading of the chapter on 'The Struggle for Existence.'*

A creationist backlash is already afoot among moderate denominations. Today 89.6% of Americans in the twelve largest Christian denominations belong to churches that support the teaching of evolution.[6] In a recently televised sermon titled "Will Teaching our Children Evolution Make Them Atheists?" the minister lamented that ignoring the knowledge of science will turn off and tune out some of our brightest minds from organized religion. There is reason to believe it already has.

Tom A. is a forty-eight-year-old engineer who likes to tinker with antique cars. His children attend a public school targeted by the creationist movement and it's been very much on his mind.

I'm not in favor of teaching both evolution and creationism in our schools. I favor teaching only evolution. Those who want creation taught have a religious agenda that they are selling. They are saying that to believe in God you must believe in creation. To believe in evolution you must be an atheist.

These people limit God by stating that He can only create people but not the process of evolution. They never consider the third option of believing in God and that He created the process of evolution. They don't realize that God created science. They charge man with inventing science. In fact, man only discovers what God created. The characterization of God as a great righteous preacher must be very distasteful to God. Someday religion, science, and evolution will be unified and the scope of God's presence will be revealed.

Strict Creationists would agree with the five key statements in Figure 9.1 that reflect their belief and show a high degree of correlation.

These same five statements comprise a "creationist belief scale" used by Frank Harrold and Raymond Eve in their classic study of belief among college students. The more statements a survey participant agrees with, the higher they are ranked on the scale of creationist belief. Do you agree with all five creationist beliefs? If so, you are a True Believer.

CREATIONIST MEGA QUESTION

Strict Creationists would agree with all five of the following:

- Everything written in the Bible is literally true.
- God created humanity much in its present form within the last 10,000 years or so.
- Adam and Eve, the first human beings, were created by God.
- There is a good deal of scientific evidence against evolution and in favor of the Bible's account of Creation.
- The Bible's account of creation should be taught in public schools as an explanation of origins.

Fig. 9-1

Only 5% of my Belief Survey participants agreed with all five creationist statements. On the other hand, over 26% of respondents agreed with at least one or more, but not all five, creationist statements. For these folks and many others the connection between science and spirituality is far more intimate and complex than suggested by standard "yes" or "no" survey questions. These results hint at a far more interesting voice in the evolution/creation debate that we'll explore next.

NOTES

1. George Bishop, "The Religious Worldview and American Beliefs About Human Origins," *The Public Perspective*, Vol. 9, no. 5, The Roper Center, University of Connecticut, 1998, p. 40.
2. Surveys by the Gallup Organization, 1991, 1993, and 1997 in Bishop, see above.
3. George Gallup, Jr. and Robert Bezilla, "Conservatives, liberals locked in ongoing biblical debate," *Sarasota Herald-Tribune*, April 20, 1996, p. 4E.
4. Raymond Eve and Francis Harrold, "Scientific Creationism and the Politics of Lifestyle Concern in the United States," in Bronislaw Misztal and Anson Shupe, eds., *Religion and Politics in Comparative Perspective*, Praeger, 1992, p. 109.

5. James Glanz, "Poll Finds Support is Strong for Teaching 2 Origin Theories," *The New York Times*, March 11, 2000.
6. Kendrick Frazier, "Science and Religion: Conflicting or Complementary?" *Skeptical Inquirer*, Special Issue on Science and Religion, CSICOP, Amherst, NY, 1999, p. 19.

REFERENCES

Davis, Percival and Dean Kenyon. 1993. *Of Pandas and People: The Central Question of Biological Origins*. Dallas: Haughton.

Harrold, Francis B. and Raymond A. Eve (eds.). 1987. *Cult Archaeology and Creationism: Understanding Pseudoscientific Beliefs About the Past*. Iowa City: University of Iowa.

Eve, Raymond and Francis Harrold. 1992. "Scientific Creationism and the Politics of Lifestyle Concern in the United States." *Religion and Politics in Comparative Perspective*. Bronislaw Misztal and Anson Shupe (eds.). New York: Praeger.

Eve, Raymond and Francis Harrold. 1993. "The Influence of Group Process on Pseudoscientific Belief: 'Knowledge Industries' and the Legitimation of Threatened Worldviews." *Advances in Group Processes* 10:133–162.

Ewing, Tanya. 1997. "Evolution? Noah Way." *Sydney Morning Herald*. March 24.

Frazier, Kendrick. 1999. "Science and Religion: Conflicting or Complementary?" *Skeptical Inquirer*. Special Issue on Science and Religion. Amherst: CSICOP.

Glanz, James. 2000. "Poll Finds Support is Strong for Teaching 2 Origin Theories." *The New York Times*. March 11.

Jones, Steve. 2000. Darwin's Ghost: The Origin of Species Updated. New York: Random House.

Kitcher, Philip. 1992. *Abusing Science: The Case Against Creationism*. Cambridge: MIT.

Numbers, Ronald. 1992. *The Creationists*. Berkeley: University of California.

Ratloff, Janet. 1996. "When Science and Beliefs Collide." *Science News* 149:360.

Whitcomb, John Jr. and Henry Morris. 1961. *The Genesis Flood: The Biblical Record and Its Scientific Implications*. Philadelphia: Presbyterian and Reformed Publishing.

WEBSITES

Teaching About Evolution and the Nature of Science (http://bob.nap.edu/readingroom/books/evolution98)
Award-winning site from the National Academy of Science. Excellent treatment of evolution, the scientific process, and creationism. Easy to understand, outstanding for teachers, but recommended for all. Instructional material and references.

Science and Creationism (http://www.nap.edu/catalog/6024html)
Newly revised report from the National Academy of Science is a companion volume to *Teaching About Evolution and the Nature of Science*. Covers the origin of the Universe, Earth, and life; evidence supporting evolution; refutes intelligent design. Can also be ordered: 1-800-624-6242.

FURTHER READING

Ronald Numbers, *The Creationists: The Evolution of Scientific Creationism*, University of California Press, Berkeley, 1992. Thorough and intriguing history of the American creationist movement.

Christopher Toumey, *God's Own Scientists: Creationists in a Secular World*, Rutgers University Press, New Brunswick, NJ, 1994. Creationism as a belief system upholding moral authority viewed through the lens an anthropologist who studied local creationist groups and interviewed their leaders over a period of five years.

Chapter 10

Room For a Different View

The test of a first rate intelligence is the ability to hold two opposed ideas in the mind at the same time, and still retain the ability to function.

F. Scott Fitzgerald, *The Crack-up*

We all like a sure thing. We like certainty and a sense of control. Simple answers to complex questions. Knowing that something is either one way or the other. Either-or is the theme song of the creation/evolution debate. For example, the creationists argue, "Either science can show us every fossil intermediate of evolution, or creation is true."

We know that strict Creationists deny that life evolved. They believe in the literal truth of the story of Creation as told in the book of Genesis and that God created humans in their present form within the last ten thousand years.

But who are the Evolutionists? Gallup polls continue to report that only 10% of Americans identify strongly with the statement, "Humans have developed over millions of years from less advanced forms of life. God had no part in this process."[1] These are the Evolutionists. They believe that all life, including humans, evolved solely through natural laws.

We identified strict Creationists with belief in five key statements concerning evolution and human origins. Similarly, staunch Evolutionists would agree with all five statements shown in Figure 10.1

EVOLUTIONIST MEGA QUESTION

Strict Evolutionists would
agree with all five of the following:

- The Earth is between 4 and 5 billion years old.
- The theory of evolution should be taught in public schools as an explanation of human origins.
- Human beings biologically identical to us today have been around for at least 40,000 years.
- Human beings came to be through evolution, which occurred without the help of God.
- Life began as a result of natural chemical reactions, and not through the power of a supernatural being or Divine Intelligence.

Fig. 10-1

Evolutionists are more likely to be found among college graduates, men, the young (under 29 years of age), and those from Eastern and Western states. National surveys show that religion is unimportant to the Evolutionists. Most state they have no religion, never attend church, and believe that the Bible is a book of fables, legends, and history.

The profile of participants in my Belief Survey is strikingly similar to that of the Evolutionists: young (two-thirds are 29 years old and under), well-educated (46% have some college), more men (57%) than women (43%), and white (81%). Nearly one out of four participants agree with all five statements of the Evolutionist Mega Question—over twice the national average of strong evolutionists. What about the others?

Is it possible to believe in human evolution and a spiritual cause for human origins?

National polling continues to confirm that 40% of Americans, or over 100 million people, identify with the statement, "Human beings have developed over millions of years from less advanced forms of life, but God guided this process, including the creation of humans."[2]

In fact, nearly two-thirds of my evolution-oriented Survey respondents agree with the statement, "One can believe in scientific evolution and the existence of God or a Supreme Being—one belief does not exclude the other." When we eliminate the strict Evolutionists and those with some degree of creationist belief from my Survey respondents, we are left with over 44% who hold a different view. This is a slightly larger proportion than national polls report and may be the result of a decidedly younger, more educated, male, and white group of Survey respondents than the national profile.

These are the Moderates. They rise above the either-or message of the creationists and evolutionists to allow room for a different view where theories of science and convictions of faith coexist. The voice of this unrecognized multitude is seldom heard above the din of the evolution/creation debate.

> Jane S. is a retired chemistry teacher. Forty years of performing standard test tube experiments have given her a unique feel for the power and potential of nature. But her beliefs about where we came from are her own recipe.
>
> *When I look at my past experiments and what I've learned about the Universe, to me it can't be an accident. There is some organizing, devising force. We use "god" or "creator" for lack of a better word, but I think of it as an experiment because I love experiments. Where or when this creator, figuratively speaking, puts in a finger or takes one back I have no idea.*
>
> *If there were convincing evidence that the soul can occur naturally, I would accept it. But I know we would continue to learn and ask the question "How did it start?" It still would not jar my belief in an overall intelligence. Behind the Big Bang, "somebody" had to punch the button.*

Moderates like Jane constitute a sizeable portion of the population, and there is evidence that their numbers may be significantly underreported. A March 2000 poll conducted by Daniel Yankelovich for the People for the American Way Foundation reported that a whopping 68% of Americans agree that it is possible to accept evolution and also believe that God's hand guided human origins. Only a third of the poll's respondents defined creationism as belief in the literal truth of the Creation in Genesis. The vast majority viewed it more loosely in terms of God creating humans, but not necessarily the way the Bible describes it. Despite inherent contradictions, these people are comfortable with a belief that mixes science and spirituality.

Moderates may be the largest group of believers in America today, where a rise in spirituality in an age of the Information Superhighway hints at a basic need to unify our physical and spiritual lives.

Most belief surveys ask questions based on the tradition of either-or-ism: "Do you believe in evolution?" or "Do you believe in creation?" Evidence from my Belief Survey and other studies suggests that this forced dichotomy leads to false choices. Moderates have no place to land and be counted. Their responses may label them "creationist" or "evolutionist," but they hold a much more complex view of human origins.

For example, in a recent survey of college students' beliefs, only 20% said they believe evolution is valid. Over 60% said they believe in creationism. But follow-up questioning revealed that many students equated creationism with a general belief in God or a higher power. They had no problem with belief in evolution <u>and</u> an ultimate cause.

Who are the Moderates and what do they believe? How do they hold two opposing ideas? Are you a Moderate with a different view?

The Long Road to a Different View

Moderates are risk takers who have embarked on a perilous psychological journey. Their worldview is a radical departure from that of the creationists. They rely on contemporary insight, not outdated tradition, to solve problems and answer moral questions. They believe that truth and knowledge are the result of human endeavor, not divine revelation. In many respects Moderates personify the intellectual inheritance of the Enlightenment. They seek knowledge through theory, testing, and empirical observation. Their sense of the spiritual is a result of this same search and contemplation, not scriptural dogma and church teaching.

Nationwide studies report that Moderates are more likely to be men than women, forty-four years of age or under, and college-educated. Unlike creationists and evolutionists, Moderates are not concentrated in one or two regions but are fairly evenly distributed throughout America.

Moderates consider religion "fairly important" in their lives, but support the separation of religion from politics and education. They are generally tolerant of the lifestyles and beliefs of others. They attend church once a month on average and are more likely to be main-line Catholics and Protestants. Moderates are far less likely to believe in scriptural truth. They might believe the Bible is "the inspired word of God" and a metaphorical, not literal, code of ethics.

Moderates are more likely to think that:

- Life is not predestined. We have choices.
- Quiet skepticism is healthy.
- Truth is acquired through questioning and staying informed.
- Beliefs should change when there is new evidence and reasoned proof.

Moderate beliefs don't spring from a "need for comfort" or "fear of uncertainty"—the tired old cliches of religious psychology. After all, what's comforting about forsaking the Rock of Ages to cling to a belief system built on certainties that can shift at any moment?

So what exactly do Moderates believe?

Not surprisingly, Moderates believe that the Earth is billions of years old, that humans evolved by means of the natural process of evolution, and that we should teach evolution as an explanation for human origins. They would not agree with any of the five key creationist questions, and at first blush might appear to be strict Evolutionists.

But then things really start to get interesting. The Moderates' openness to spirituality first emerged in my Survey and in interviews when the questions touched on spiritual intervention in the detailed process of human evolution and the beginning of life. For example,

- Moderates accept the evolution of humans, but not all agree that "Human beings came to be through evolution, which occurred without the help of God."
- They also give mixed responses to the statement: "Life began as a result of natural chemical reactions, and not through the power of a supernatural being or Divine Intelligence."
- Most agree that life itself was a natural chemical evolution, and that life could easily have evolved on other planets. But for some, the spark of life itself remains open to spiritual interpretation.
- Many Moderates are virtually indistinguishable from strict evolutionists, but leave room for an Ultimate cause that set the physical laws of the Universe in motion.
- Most Moderates describe themselves as "not religious," but many express doubts about whether the Universe and life itself could have come about without a guiding hand for which they have no name.

Some Moderates are simply searching for answers to key questions—"Does the soul exist?" or "What happens when we die?" or "What existed before the Big Bang?" They reluctantly acknowledge that they use words like "God" or "Higher Power" to fill in the blanks where modern science is ignorant.

Moderate beliefs are more heterogeneous than those of creationists or evolutionists. The nuance of this sliding scale of Moderate spirituality is captured in their personal expressions of belief.

> Mary G., a widow in her sixties, raises prize-winning orchids. She never studied evolution in college, but developing new species of plants has given her an up-close-and-personal experience with change.
>
> *I believe God set up the world and that it has changed since then. Species have evolved and changed through natural laws. God set up the laws of evolution, so He's still in control to the extent that he set up how things would change.*
>
> *I think that God's intervention was pre-life. I think God created the first atoms and the atomic theory, and survival of the fittest, and all the things that over time we have discovered about the world and how it works. I don't think the Bible is necessarily wrong, but I don't believe in it word for word. I think there are general principles in it to follow. But if science does find out something, then it is fact.*

For many Moderates like Mary, evolution is simply the story of how life changed after God created the laws of nature.

Above all, Moderates are learners. The Moderate personality tends to be open, curious, and optimistic. They are independent thinkers who often develop atypical ways of looking at science and belief. In his book *Skeptics and True Believers*, Chet Raymo defines the True Believers, best exemplified by the creationists, as "yearners" and the Skeptics as "learners." But Moderates are skeptical learners who yearn to know more about those questions that science can't answer. They generally believe that the inexplicable will be explained, but recognize that the answers may involve dimensions of reality beyond the domain of science.

Throughout history people have maintained a sense of the spiritual in an increasingly scientific world. The philosophy known as *deism*, rooted in 17th century France and prevalent in England during Darwin's time, melded scientific theory and religious tenets into non-creationist beliefs about human origins. Deists believed evolution was the inevitable result of natural laws put in play by God. Many deists believed in God's intimate meddling in daily life. But they were soon outnumbered by other deists who claimed that a Creator would take no interest in what had been created. Some deists accepted the moral teachings of the Bible, but most rejected scriptural truth and renounced affiliation with any specific religious denomination. A few denied the existence of any supernatural order.

By the end of the 1700s the popularity of deism had spread to America's upper-class. George Washington, Thomas Jefferson, and John Adams were deists. Their words "one nation, under God" are perhaps best understood in the light of their deistic convictions. If all this sounds familiar, it is. Today many of our Founding Fathers would be considered Moderates.

Deist is an outdated term that overlooks the enormous transformation in science and religion since the days of Newton, Washington, and Darwin. Assorted shades of spirituality thrive in our modern culture that would not have been imaginable as recently as fifty years ago. Religion is no longer place-oriented, rooted in a physical church or close-knit community. Modern worship is multifaceted, often combining orthodox ritual with non-traditional practices such as chanting and meditation. Many of us draw our sense of the spiritual from inward reflection and long walks on the beach.

Philosopher William James claimed that we believe what works for us. Today's believers are practice-oriented. They borrow bits and pieces from various creeds to guide their actions and behavior. Moderates often take this tendency to the maximum. They must hold on to the beliefs that organize their lives even when scientific certainty isn't available. In addition they recognize that beliefs must change as new information becomes available. Abandoning an outdated belief is considered healthy and wise, not heretical.

Moderates are rarely invited into the evolution/creation debate. Neither side wants to claim them. Creationists shun Moderates as atheists and heretics whose religion is the godless materiality of evolution. Evolutionists dismiss the Moderates as step-children of the fundamentalists and toss them back into the camp of the creationists.

It's clear that Moderate beliefs are fodder for the creationist's zeal to uphold the reality of the unseen. But it's more difficult to fathom the reaction of the evolutionists. Moderates would prove to be loyal allies on the side of science. In fact, they could place a foot in both camps and enhance the prospects of true conciliation. But their reasoned voice is muted by scientists fearing any hint of doubt that science alone can explain and predict the material world. And as biochemist Erwin Chargraff admitted, "Science is wonderfully equipped to answer the question 'How?' but it gets terribly confused when you ask the question 'Why?'"

Just Say No

In the minds of atheists, "why" is irrelevant. They just say no to the reality of the unseen. The number of professed atheists among Americans rarely rises above 2%. But global surveys of religious belief report that over 900 million people, or 15% of the world's population, consider themselves atheists or non-religious. Over 16% of my Survey participants described themselves as atheists. Journalist H. L Mencken summed up the position of his fellow atheists when he wrote, "For centuries theologians have been explaining the unknowable in terms of the-not-worth-knowing."

Atheists and creationists have one very important characteristic in common. Both take an extreme position in the evolution debate and argue from an either-or point of view. Strict creationists would regard all evolutionists as atheists. Many are.

Evolutionist and avowed atheist Richard Dawkins dispenses with any personal need for spiritual belief by saying that in his view science not only answers the "how" but the "why." He explains the alleged healing power of prayer as simply evidence for the placebo effect. He believes that "why" isn't a legitimate question to ask about a physical Universe that has no purpose and is not the result of design or intent.

> Bob T. is a 54-year-old mechanic whose hobby is building simple robots. He is fascinated by high-tech gadgetry and believes that humans will survive because of our down-to-earth pragmatism—not belief in the unknown.
>
> *It wouldn't be possible for me to believe in evolution and God. I worked for some time with people in science. It seemed they kept a wall between their work and going to church on Sunday. One had nothing to do with the other.*
>
> *My father taught me to think things through to the end. Only then can you have your own answers. My answer is that you can live a moral and ethical life without having a formal religion.*
>
> *I believe that man has a spirit in him that the other animals don't have. Not a soul but special talents and motivation. I don't believe in the soul. You're talking to a total and complete atheist.*

If atheists say "No," then agnostics say "I Don't Know." They doubt that the existence of a transcendent being can ever be proven. Agnostics don't have a sure answer to a lot of life's questions, but are open to finding out. They not only ask "why?" but sometimes "why not?"

> Elizabeth M. is an administrative assistant to the president of a small industrial firm. College-educated, single, and in her thirties, Elizabeth retains a marked independence both in business decisions and personal convictions.
>
> *The world seems so vast with all the planets and geological eras. I think it's unknowable. You tend to think that someone or something invented it all. Evolution seems the most likely explanation, but it doesn't tell us how life began. That's where our knowledge stops.*

> *I call myself an agnostic. I'm not an atheist or anti-God. I don't know how others can be so sure. Religion came from the dreaming of primitive people, a need to explain why bad things happen to good people. A lot of people search for help to get through life and find it in religion. I've had few misfortunes and haven't had to find reasons.*

Despite appearing skeptical and wishy-washy, most agnostics like Elizabeth would strongly agree with Clarence Darrow, "I do not consider it an insult, but rather a compliment to be called an agnostic. I do not pretend to know where many ignorant men are sure."

Darwin professed to be an agnostic, as did many of his era who believed life evolved without divine intervention or a predetermined path. It is probably safe to say that most atheists and agnostics accept the theory of evolution as the most reasonable cause for human origins.

Where the Rubber Meets the Road

Any hint that the rift between science and religion is closing makes headlines. But recent efforts to warm the chilly detente between the two have only emphasized the lack of a common language and frame of reference.

On October 25, 1996 the world awoke to news that Pope John Paul II had endorsed evolution—as part of God's master plan. *The New York Times* enthused, "Pope Bolsters Church's Support for Scientific View of Evolution." Italy's conservative *Il Giornale* chided, "Pope Says We May Descend from Monkeys." Appalled creationists in Kentucky countered, "The Pope's assertion directly contradicts the Bible. It can't be compatible with the Christian position."

In California, The National Center for Science Education, a group that monitors threats to the teaching of evolution in schools, praised the Pope's remarks as "a very clear statement that there is no conflict" between science and religion. Catholic educators also took it in stride, saying "We always start with the premise that all creation came from God. Beyond that it's a scientific issue." Four years earlier the Pope had apologized for the Roman Catholic Church's denunciation of Galileo. Suddenly reconciliation was in vogue. Both the National Academy of Sciences and the American Association for the Advancement of Science, publisher of the prestigious journal *Science*, launched programs to promote conciliatory discussions.

The Templeton Foundation, dedicated to the promotion of religion and spiritual values in America, sponsored the "Program of Dialogue Between Science and Religion." Its Conference on Contemporary Cosmology and Religion brought together over two dozen experts, from Nobel laureates to leading-edge theologians, in an effort to find common ground at the crossroads between science and religion. But a glance at chapter headings in the publication *An Evolving Dialogue: Scientific, Historical, Philosophical and Theological Perspectives on Evolution* reveals few areas of overlapping interest.

Nevertheless hundreds of new organizations are sprouting to foster a creative exchange of views on science and religion. Seminaries have formed study groups to bridge the gap. From Chicago to Berkeley, and Princeton to Cambridge, ministers preach the moral message of the Big Bang. Believing physicists come out of the closet to publicly share their opinion that the background radiation spawned by the Big Bang represents "the handwriting of God." Science can't spy God at the end of a telescope, but some scientists claim that learning about the Universe can provide clues to the character of the Divine. These outings are still viewed as fringe activity by mainstream science.

But not all scientists are atheists and not all belief is irrational. Many scientists would certainly be considered Moderate believers. As one former researcher commented, "I don't know how a scientist can be a confirmed atheist. If the world emerged from chaos, why has it lasted so long?"

How We Reconcile Science and Faith

It all comes down to questions of science versus questions of faith. Or as Pope John Paul II put it, "The sciences of observation describe and measure the multiple manifestations of life with increasing precision, and correlate them with the timeline. The moment of transition to the spiritual cannot be the object of this kind of observation." In other words, science can tell us how we evolved from monkeys, but it has nothing to say on the subject of the soul.

Unfortunately, many scientists blur the distinction between science and the soul or spirituality when they concoct phrases such as "God particles" and "the mind of God." George Coyne, Jesuit priest and director of the Vatican Observatory, objects to scientists who use "God" as a catch-all for unexplainable phenomenon, or cite science to prove God's existence. Coyne claims, "There is no scientific teaching in Genesis. The understanding of origins has nothing to do with the existence of God, but it has a lot to do with my understanding of God."

The late evolutionist Stephen Jay Gould agreed. In his book, *Rocks of Ages: Science and Religion in the Fullness of Life*, Gould interpreted the Pope's comments as clear acknowledgment that the investigative and theoretical areas of science and religion don't overlap. Science seeks answers to "what" and "how" by gathering facts and developing theories. Religion seeks answers to "why" by interpreting scripture to grasp the morality and meaning of life. Science and religion operate in parallel worlds and should stay out of each others sandbox. However, this separate-but-equal view of science and religion ignores the subtle communication between the two that takes place every day in the minds of Moderates, including many scientists.

Another scenario envisions science and religion working as partners—each advancing the other's goals by investigating questions typically outside their normal arena. What could those questions possibly be? Will science investigate the risen Christ as a phenomenon of atomic wave theory? So far any accommodation between the two disciplines has been one-way. For example, theologians have attempted to reinterpret the doctrine of original sin in light of the evolution of humans from apes. But I'm not aware of any evolutionist who has tried to identify the species of serpent responsible for Adam and Eve's downfall.

Some scientists think they have found the answer to "why" in their notion that intelligence determines the nature of the Universe. They believe that we can study and measure the Universe only because its physical laws are compatible with life and can be observed by intelligent beings. Were the laws of the Universe fine-tuned just for us? It's an interesting piece of circular logic that is fairly obvious. But should we extend this principle to all life?

The conclusion of these scientists is that human beings were the predetermined goal of the creation of the Universe. The Big Bang was intentional.

The conclusion of most scientists? Absurd! Slight deviations in initial conditions would have created other kinds of Universes where life, in some form, could have evolved. The initial conditions we observe may be necessary for life as we know it, but not sufficient. But what existed before the Big Bang?

John J. is an archaeologist and astronomer. He has traveled the globe from Mexico to Europe, analyzing manmade celestial calendars such as the stone circle at Stonehenge. John shares the deep reverence of ancient cultures for the power of the Universe, and has created his own very personal cosmology.

It's impossible to know the mind or creation techniques of God. But if you say God was not active at the beginning of time, you deny God. He proclaimed both natural and spiritual laws. He interacts with the Universe and humans within the boundaries of the laws. The Universe, evolution, and life can only exist with the presence of God.

The panorama of Moderate beliefs and the psychological concessions they reveal are as far reaching and multi-dimensional as the human imagination.

How do Moderates reconcile science and faith? I don't believe they do, at least not in the ordinary sense of reconciliation. There is no definable nexus where the facts and laws of science happily nestle in the clutches of faith and dogma. No seamless segue—only that very personal and uniquely individual comfort zone where the two bump and collide gently, coexisting with a minimum of cognitive disruption. The difference is this. Moderates—unlike strict evolutionists or creationists—take the separate-but-equal relationship between science and faith to a new level and welcome a constant feedback between the two. One area of knowledge constantly informs the other. The strength of the Moderates' view is their lack of a need for certainty. Moderate beliefs thrive in a sphere of uncertainty where spirituality can and must grow and expand with the knowledge of science. Things don't always have to make logical sense.

The beliefs of Moderates express our search for a deeper meaning in the picture painted by science. During the past decade, studies that gauge the nature and extent of religious belief in America have shown a marked rise in this search. In the final analysis, it's all about cause. We want to know why there is something instead of nothing.

NOTES

1. George Bishop, "The Religious Worldview and American Beliefs About Human Origins," *The Public Perspective*, Vol. 9, no. 5, The Roper Center, University of Connecticut, 1998, p. 40.
2. See Bishop, above.

REFERENCES

Easterbrook, Gregg. 1997. "Science and God: A Warming Trend?" *Science* 277:890.
Einstein, Albert. 1950. "Science and Religion," *Out of My Later Years*. New York: Philosophical Library.

Eve, Raymond and Francis Harrold. 1993. "The Influence of Group Process on Pseudoscientific Belief: 'Knowledge Industries' and the Legitimation of Threatened Worldviews." *Advances in Group Processes* 10:156.
Glanz, James. 1999. "Celebrating a Century of Physics." *Science* 284:35.
Gould, Stephen Jay. 1997. "This View of Life." *Natural History* 106:16.
__________. 1999. *Rocks of Ages: Science and Religion in the Fullness of Life.* New York: Ballantine.
Marklein, Mary Beth. 1995. "Pope accepts evolution, creates furor." *USA Today* October 25.
Miller, James B. (ed.). 1998. *An Evolving Dialogue: Scientific, Historical, Philosophical and Theological Perspectives on Evolution.* Washington, D.C.: American Association for the Advancement of Science, 1998.
Seife, Charles. 2001. "Science and Religion Advance Together at Pontifical Academy" *Science* 291:1472.
Whitehead, A. N. 1947. *Science and the Modern World.* New York: Macmillan.

Chapter 11

What Is the Answer?

I hate quotations. Tell me what you know.

Ralph Waldo Emerson

Here's the truth as I see it.

I. **All the hard evidence that has come out of the ground, or has been extracted from test tubes, tells us that <u>life has evolved naturally</u>.**

Gaps in the fossil record? What gaps? Detailed fossil sequences in thousands of animals demonstrate the obvious steps of evolution and common descent.

- Fossil and molecular evidence clearly proves that whales were once four-legged landlubbers related to hippos that decided to take a swim—and never looked back. Their bones and genes retain vestiges of that evolutionary journey.

Who says evolution didn't happen because no one saw it? Today researchers are watching and documenting evolution.

- Many organisms—fruit flies, sockeye salmon, insects, and others in the lab and in the wild, can be observed as they develop traits that lead to the emergence of new species.

Not only is all life connected by chemistry, but it also bears the deep scars of parallel development. We see similar patterns of form and function in limb and lung that have been constructed, torn down, and reused over millions of years.

- The close parallel among evolutionary pathways of all types of eyes has been confirmed with discovery of the *pax6* gene, the universal instruction manual for making eyes in all animals. Dissimilar complex organs, such as the eyes of flies and humans, share the same evolutionary pathway.

Don't listen to the Intelligent Design crowd either. Complexity in evolution *can* be reduced to its smallest components and explained using well-known chemistry and physics. You just have to look at the evidence and believe the facts.

- Blocks of developmental genes are strikingly similar in very different animals. Parallels in their action and organization promote limb and rib development in mouse, chick, and snake embryos. Evolution is not only clever, but efficient.

The hidden sources of genetic variation needed to evolve complex organs and features has been found in the accumulated wisdom of our genes. The assault of millions of years of evolutionary pressure has sorted, rearranged, and fine tuned them into efficient machines that act quickly and in concert to produce innovative change. Nature has been the (intelligent) designer, courted by the whims of Evolution—a purposeless, directionless, and unlimited force in our world.

Some evolutionists translate Darwin's *Origin of Species* like sacred text. Others delight in tackling exceptions to the rule, expanding and updating the details with 21st century knowledge. Their vigorous challenges have caught the attention of creationists, who hang on every word like media junkies at election time looking for a negative soundbite. But the scientific dissension is normal and healthy. It's the way science works—questioning, testing, and expanding the evidence and filling in the details.

II. Evidence and sound reasoning point to natural evolution as the architect of all life on Earth—including the evolution of modern humans.

Hold-outs for an answer other than a natural cause for human origins will have a bumpy ride as science continues to peel back the onion of evolution.

- In 2001 and 2002 fossil finds expanded our knowledge of human evolution three million years further back in time. Fossil sequences are rapidly filling in the picture from Darwin's "missing link" to us.

- Chimps and humans differ genetically by less than 2%. But this small genetic difference belies striking differences in levels of *gene activity* in the human brain compared to that of chimps. In all other organs gene activity is similar, but genes in the human brain are working at higher levels of protein output—indicating different operational dynamics.

- Overall *brain activity* in chimps and modern humans also differs in several "hot spots." These spots correspond with the same areas on the surface of the right side of our brain that show the most differential physical expansion in fossil skulls of human ancestors throughout 2.5 million years of evolution. These areas of heightened activity help us process the subtleties of language—a key factor in that decisive change in culture and behavior known as the Great Leap Forward that appeared with the arrival thirty-five thousand years ago of fully modern humans.

The emergence of modern humans was an arduous journey, not a miraculous event. We would have been known as just another ape except for three pivotal forks along our evolutionary trail:

1. Four to six million years ago our ancestors got into the habit of walking upright, freeing their arms and hands to eat and fight.

2. Three million years ago our line split into two distinct body types (robust versus lightly built) who ate different types of foods (veggies versus meat) and occupied different ecological niches.

3. By 2.5 million years ago the slimmer guys and gals were routinely making and using tools and becoming us, while our robust cousins munched on to an evolutionary dead end.

Over two million years later we arrived as fully modern humans through short-term, pragmatic steps made possible by the dynamics of natural selection and survival of the winning physical and mental designs. As Darwin described it in *The Descent of Man*,

> *We have given to man a pedigree of prodigious length, but not, it may be said, of noble quality. The world appears as if it had long been preparing for the advent of man: and this, in one sense is strictly true, for he owes his birth to a long line of progenitors. If any single link in this chain had never existed, man would not have been exactly what he is now.*

The trail of human evolution, drawn by scientists who make the stones and bones speak, relies on the same laws of chemistry and physics that govern the Universe. But many people eagerly suspend these laws when it comes to their personal beliefs about human origins.

As a species we carry a lot of baggage from millions of years of evolutionary struggle. The price we have paid for survival is a complex suite of inherited instincts that rely on very powerful passions. These same instincts mold our ego, resulting in an emotional intellect that can rationalize anything. It defends our ego against threats to our place at the center of the natural world.

Darwin insulted the accumulated ego of our species, and many of us instinctively deny, argue, and rationalize away the hurt. This is the emotional agenda of creationists and promoters of Intelligent Design.

The key difference between theories of science and dogma of faith is that science has a built-in escape valve that guarantees new theories won't be suppressed if evidence corroborates them. This is exactly what is happening today in the study of evolution. As scientists forge a new evolutionary paradigm for the 21st century, will the processes they uncover convince us? Or will we remain skeptics until we see evolution with our own eyes?

III. And what about the soul, you ask?

Well, what about it? Immortality dies hard. Believers in the reality of an unseen dimension rely on hope. And I hope they're right. There is not a single person, not even an atheist, who wouldn't jump at the chance to be reunited with those we've loved and lost. But if I had to place a Las Vegas bet on it, I wouldn't play.

Past life recall, déja vu, sightings of astral bodies, and channeling are a mix of checkable facts and fantasies, not proof of a soul or reincarnation. There is not a shred of proof that we can tap into paranormal powers or another dimension of reality. No one could keep a secret like that. If someone really had that kind of knowledge they would become rulers of nations, not call takers on the psychic hotline.

Perhaps the most evidence I can offer for a life beyond this one was expressed by a friend who admitted, "I don't believe in reincarnation. I believe we live on in the minds and deeds of those we leave behind."

Is the notion of a soul simply our highly developed consciousness-with-a-conscience? I believe it is. I believe that the most fascinating pursuit of modern science is research into human consciousness. As we develop better techniques to probe the active human brain, many things like the "mysteries of the soul" will be revealed as wondrous and explainable consequences of being human. And wouldn't *that* be awesome?

IV. Funny thing about the origin of life.

Like contemplating the size of the Universe or the shape of infinity, our mind goes fuzzy straining to grasp our beginnings.

Wherever and whenever life began, it was an evolutionary process. Life on Earth probably began not once, but many times, only to be destroyed by the same energetic forces that made it possible. Natural selection and evolutionary success on Earth may even have favored imported chemistry and coded information from other worlds. Imagine a bush sprouting many kinds of replicating molecules and many primitive cells, each based on a slightly different chemistry. Only the best adapted seeds of life survived cosmic bombardment and lethal environments and passed their survival strategies on to their offspring.

The point isn't whether Earth is unique. Every planet we come across—and we're finding more every year—will be unique. The question is whether Earth's environment and place in the Universe is *essential* for complex life. There is no evidence to suggest that this is the case. Everywhere we look we find life as we never knew it. In the chaos of undersea volcanic vents, inside hot sulfurous springs, and in solid rock buried miles beneath the planet's surface, we find life—consuming, digesting, reproducing, changing, and evolving.

Our picture of life is biased and certainly naive. Life on other planets may rely on a completely different chemistry from the one we know. But there is good reason to believe that on other planets located around other stars where there is water and an abundant energy source, there will be simple life at the very minimum. Complex life takes longer to evolve and may be more rare. Other planets might have lots of bacteria, but no elephants or palm trees.

Our ego whispers that we are the only intelligent beings in the Universe, that Earth was made for us alone. It's a heady thought. But the circular logic of the theory that we are able to study the

Universe only because its laws were fine-tuned from the very beginning to allow for our evolution is not only characteristic of irrational belief, but it's also bad science. It's arrogant to conclude that we are the reason for the happy coincidence of physical forces that have allowed life to develop on Earth.

Beyond our lonely galaxy there are far older reaches of the Universe where time and alien environments may have spawned sentient, intelligent life. The discovery of conscious alien life could spell danger to beliefs bound to the written word. Any accommodation would be awkward and fragile. The psychological underpinnings of many believers would be temporarily swept asunder, until the life raft of rationalization provided their salvation. Would an intelligent alien have a soul? Some theologians might rationalize that life on other planets simply expands the missionary opportunity. I, for one, believe that our biggest challenge will be to let them know that we come in peace.

V. Are we still evolving?

Technically yes, because our gene pool is still changing. Despite our much touted ability to adapt through technological innovation, Darwin's concepts of change and natural selection are still alive and well. What about genetic engineering and cloning? Won't these help us master the survival game? Let's look at that possibility.

We are sexual animals. Sex has made all the difference because it masks the bad effects of genetic mutation. It allows unique, beneficial combinations of genes to emerge, and stockpiles variation for the future. Each offspring is (genetically) more than the sum of its parents.

Cloning robs us of the genetic advantages of sex. A cloned cell produces others just like it that carry the same genetic profile. Clones have all their parent's flaws but none of the advantages of new combinations of genes. In the process of cloning, some new mutations will occur that won't be buffered by the addition of genes from a second parent. Mutations that would be viable and even beneficial in sexual offspring, will never see the light of day.

Cloning robs a species of the genetic variation that is the raw material of evolution.

Genetic engineering shows great promise for curing disease and redesigning immune systems. No more cancer or sickle cell anemia. But unknown benefits that one hundred thousand years of evolution have built into the modern human genome could be destroyed by innocent fiddling. For example, tampering with key immune genes could cure cancer but frustrate our ability to fight off super bugs and new diseases like AIDS. Natural selection is not a charitable institution. It is indifferent to the weak, the suffering, and the genetically engineered.

What will the next one hundred thousand years years be like on Earth for modern humans? Yogi Berra reminded us that "prediction is very hard, especially when it's about the future." But let's give it a try.

In the early 1990s a team of American scientists drilled a five-inch core through two miles of ice formed from two hundred and fifty thousand years of Greenland snows. Definable bands in the ice core captured ancient climate changes like frozen fossils. In the finer details of those changes the researchers found astonishing and frightening evidence of repeated occasions when Earth's climate has changed drastically in ten years or less. Our stable climate of the past ten thousand years is a pronounced exception to those more volatile climatic swings of the past. If we were to experience a sudden ice age, we would find ourselves in the same boat as the millions of species that have come and gone on this planet, cast aside by the whims of nature. How effectively could current technology and politics mobilize to save us? Modern humans haven't changed physically for over fifty thousand years, but we have learned to change our behavior. Would our knowledge of evolution and mass extinction suggest new behaviors to save us from a global catastrophe? Or are we only a temporary life form, destined to make way for the new mammals of the next epoch? This startling possibility has far reaching implications for science, religion, and anyone's beliefs about an ultimate cause.

VI. What is the connection between science and spirituality?

Albert Einstein summed up the elusive relationship between science and religion when he wrote, "Science without religion is lame, religion without science is blind." But can they be partners?

There is an underlying frustration with the inability of science to provide final answers to many of life's most profound questions: Why does the Universe work the way it does? What existed before the Universe? Or, as one young teenager put it, "Why is there something, instead of nothing? If there was only chance, and no choice, there wouldn't be any morals. When we die, if we're dead and there's nothing else, then I should be partying all the time." These are the "why" questions that we relegate to religion—or leave unanswered.

Einstein's famous comment that "God does not throw dice" expressed his displeasure with quantum theory and the inability of its laws of chance to satisfy our need for simple answers to ultimate questions. An attempt to quell this frustration is at the core of the beliefs of many people, such as the Moderates, whose views contain a high degree of cognitive dissonance—that state of mind of someone who can comfortably rationalize two conflicting philosophies at the same time. These beliefs often appear contradictory and may be influenced by that lingering feeling first acknowledged sixty years ago by Alfred North Whitehead:

> *Scientific theory is outrunning common sense. The eighteenth century was a triumph of organized common sense. It grounded itself upon what every plain man could see with his own eyes, or with a microscope of moderate power. Today we are at the opposite pole of thought. Heaven knows what seeming nonsense may not tomorrow be demonstrated truth.*

And that's the problem. We simply don't know what we don't know. Truth is relative. It is based on the sum of accumulated knowledge. What was once the province of mythology and theology is increasingly the domain of science. The "mysteries" of past generations are being solved, such as the nature of matter and energy and the material basis of human consciousness. But new knowledge brings new questions and new "mysteries." For example, scientists are still searching for the missing matter and energy predicted by the theory of the Big Bang. They are finding it spread throughout the cosmos, organized into structures and patterns that current technology can only begin to discern. As English geneticist J.B.S. Haldane predicted, "My suspicion is that the Universe is not only queerer than we suppose, but queerer than we *can* suppose."

The truth about truth and knowledge is that it's not only relative, but personal. Socrates said that there are two types of *truth*—the *truth of science* that corresponds with a knowable reality, and the *truth of personal opinion*, or what we believe in our minds is right because it is consistent with our own ideas. And as we've seen in interviews throughout this book, these two truths constantly vie with one another in our beliefs about where we came from.

Ever since our earliest ancestors stood upright and left their first footprints in the sands of time, this interaction between the two truths of reality and belief, in all its forms, has powered human curiosity and a thirst for knowledge. It is the strength of our humanity and the key reason for the success of our species.

Huston Smith, consummate historian of human beliefs, paints a picture of this intimate connection between science and belief as we journey toward ultimate answers:

> *Reality is steeped in ineluctable mystery . . . the more we understand it, the more we become aware of additional factors relating to it that we do not understand. In mysteries, what we know, and our realization of what we do not know, proceed together. The larger the island of knowledge, the longer the shoreline of wonder.*

And that is why there will always be questions about The Human Question.

REFERENCES

Freud, Sigmund. 1923c; reprint 1962. *The Ego and the Id* . New York: W. W. Norton.

Smith, Huston. 1991. *The World's Religions: Our Great Wisdom Traditions*. New York: Harper San Francisco.

Appendix

And Our Survey Says . . .

Over the course of a year beginning in spring 1995, I conducted a Belief Survey designed to use purposive and convenience sampling techniques, and not intended to reflect the opinions or attitudes of the American population at large. Instead, my approach focused on a population of interest that expressed a desire to participate in the survey and take advantage of the opportunity to share their views. The survey also gave me access to a large number of people from whom I could solicit in-depth, personal interviews about their beliefs. I understand the limitations of this methodology, but am comfortable with employing the data as I have in this book.

Methodology

The Belief Survey Questionnaire was developed by examining previous research in this area and drawing from existing questions related to individuals' beliefs about human origins and the beginning of life. I am grateful to Francis Harrold and Raymond Eve for allowing me to borrow extensively from survey questions reported in their book, *Cult Archaeology and Creationism*.

I also incorporated newly created questions specific to the focus of this book. I am indebted to Liesel A. Ritchie, Research Scientist and Fellow, and Jay Ritchie, Research Associate, both at the Social Science Research Center, Mississippi State University, for help with survey design, questionnaire validation, and data analysis.

The questionnaire consisted of two parts. Part I included standard sociodemographic items such as age, sex, race, educational attainment, and employment status/occupation. This first section also asked respondents about their reading activities, college experiences (if any), their parents' level of education, religious affiliation and spiritual activities, and their beliefs about evolution.

Part II consisted of thirty-eight Likert Scale statements pertaining to respondents' beliefs about the origins of human life, evolution, the creation of earth, science, "pseudo-science," and other related topics.

In 1995 I began to collect survey data by distributing a pencil and paper version of the questionnaire to retirement facilities, local clubs and organizations, and other interested individuals who learned of my work. The questionnaire was then adapted for Internet distribution, and a web-based survey was posted at [*www.coolsite.com/origins.html*] for general consumption and also passed along to a number of university and public web lists.

A total of 1,056 people participated in the survey both on-line (1,003) and on paper (53). Their answers were coded and analyzed using SPSS (Statistical Package for the Social Sciences). Surveys that were returned incomplete (with substantial numbers of items left blank) were not included in the data analysis.

Who Shared Their Views?

Of the more than 1,000 individuals who responded to the survey, 57% were men and 43% were women. Because so many of the survey respondents participated via the Internet, it is worth mentioning that statistics on Internet usage from 1996 report that two-thirds or more users were men and one-third or less were women. (5th Annual World Wide Web Users Survey conducted by the Graphics, Visualization, & Usability Center [GVU]; SURVEY.NET Poll) As noted in the 1996 CommerceNet/Nielsen Internet Demographics Survey, users of the World Wide Web "are upscale, professional, and well-educated."

In my survey, most respondents (41.5%) reported they were 22 years of age or under, with an additional 24.4% between 23 and 29 years of age, 14.8% were ages 30 to 39, 15.7% were ages 40 to 59, and only 1.3% were sixty years of age or older. It does appear that participants in the Belief Survey were somewhat younger than those measured by Internet usage studies for the same time frame—likely due to the fact that my survey was posted to several university-based listings.

Over 81% of survey participants were White—close to the 82.3% reported by the only usage study (GVU) that included racial data for the same time period. Other respondents indicated they were Asian (4.0%), Hispanic (2.4), Black (1.8), Native American (0.6%), or Other (3.0%).

As might be expected of Internet survey participants, a large number (29.4%) indicated they held graduate level or professional degrees. Of note is that 46.8% of respondents had at least some college education—9.5% one year of college; 9.8% two years; 8.9% three years; and 18.6% four years of college. These percentages make even more sense when we see that a full 45.0% of those participating in the survey indicated they were students. Almost 40% of those responding to the survey reported that they worked full time.

A large majority of survey participants (80.6%) grew up in the United States. Most described the area in which they were raised as a medium/metro area (32.2%) or a small town/city (29.9%). Over 21% grew up in a large metropolitan area, while 13% grew up in a rural setting. Respondents' parents tended to be well-educated—47.2% of mothers and 54.5% of fathers had a college or professional degree. Most individuals in the survey (58.3%) read more than ten books a year. They had taken college-level courses ranging from anthropology (33.6%) to astronomy (22.3%), chemistry and physics (54.7%), history (64.8%), psychology (53.7%), biology (55.6%), geology (21.9%), logic (35.3%), archaeology (14.7) and religion (27.9%).

Perhaps the most interesting profile data concerned the information sources ranked most and least reliable. Participants' top five sources for "very reliable" information were:

1. Professional science journals (e.g., *Science* or *Nature*)
2. PBS programs (e.g., "Nova" or "Cosmos")
3. Popular science magazines *(Discover, Scientific American)*
4. College instructors
5. Parents

Sources rated "least reliable" included science fiction books, books on UFOs or the Bermuda Triangle, pseudoscientific TV programs ("In Search Of," "That's Incredible") and tabloid newspapers—55% didn't read them at all. Semi-structured Interviews were conducted with randomly chosen volunteers. Excerpts of these and other less formal discussions also appear throughout the book.

The entire survey questionnaire follows.

BELIEF SURVEY
PART I
(Please circle the letter corresponding to your answer.)

1. Your age:
   a) 22 or under
   b) 23–29
   c) 30–39
   d) 40–59
   e) 60 and over

2. Your sex:
   a) Female
   b) Male

3. Which best describes your racial background? (select one):
   a) White
   b) African-American/Black

c) Hispanic
d) Native American
e) Asian
f) Other

4. Highest level of education completed (select one):
   a) Elementary school
   b) High School
   c) Technical or trade school

   College/University:
   d) Freshman
   e) Sophomore
   f) Junior
   g) Senior
   h) Graduate or professional school (medical, law, etc.)
   i) Other

5. Area of academic major or interest (select one):
   a) Anthropology
   b) Other Social/Behavioral Sciences
   c) Natural/Physical Sciences
   d) Liberal Arts/Humanities
   e) Engineering/Computer Science/Architecture
   f) Business Administration
   g) Other ________________
   h) None

6. Are you (select one):
   a) Working full-time
   b) Working part-time
   c) Student
   d) Homemaker
   e) Unemployed
   f) Retired

7. What is/was your occupation/profession? ___________
   ____

8. How many books do you read per year (outside of class requirements)?
   a) 0–1
   b) 2–4
   c) 5–10
   d) over 10

For Questions 9–18,
please tell us whether you have taken
a college-level course in each of the following areas:

Y = YES N = NO

9.	Anthropology (except Archaeology)	Y	N
10.	Astronomy	Y	N
11.	Chemistry or Physics	Y	N
12.	History	Y	N
13.	Psychology	Y	N
14.	Archaeology	Y	N
15.	Biology	Y	N
16.	Geology	Y	N
17.	Logic	Y	N
18.	Religious Studies	Y	N

19. Mother's education (select most advanced level completed):
    a) Elementary school
    b) High School
    c) Technical or trade school
    d) College/University
    e) Graduate or professional school (medical, law, etc.)
    f) Don't know

20. Father's education (select most advanced level completed):
    a) Elementary school
    b) High School
    c) Technical or trade school
    d) College/University
    e) Graduate or professional school (medical, law, etc.)
    f) Don't know

21. Where did you grow up, mostly? (select one)
    a) In the country (rural area)
    b) In a small town/city (pop. below 50,000)
    c) In a medium-size metro area (50–500,000 people)
    d) In a large metro area (over 500,000 people)

22. In what country or area did you grow up? (select one)
    a) U.S.A.
    b) Britain
    c) Canada
    d) Latin America
    e) Middle East
    f) Africa
    g) Asia
    h) Europe (not Britain)
    i) Other ______________________________

23. If you grew up in the U.S.A., in what state was this?
    ____

24. How important is religion or spirituality in your life? (select one)
    a) Unimportant
    b) Somewhat important
    c) Very important

25. What is your religious affiliation? (select one)
    a) Roman Catholic
    b) Eastern Orthodox
    c) Protestant (e.g., Presbyterian, Methodist, Baptist, etc.)
    d) Jewish
    e) Latter-Day Saints
    f) Moslem
    g) Buddhist
    h) Hindu
    i) Other ________________________________
    j) None

26. About how often do you attend church, or a spiritually oriented group meeting/discussion, in a year?
    a) never
    b) 1–4 times
    c) 5–10 times
    d) 11–20 times
    e) 21–52 times
    f) over 52

27. Which of the following terms best describes your religious or spiritual beliefs? (select one)
    a) Fundamentalist
    b) Conservative
    c) Moderate
    d) Liberal
    e) Agnostic (You don't know if there is a God or a "higher power," but you don't deny that one may exist)

f) Atheist (You deny the existence of God or a "higher power")

28. Were you taught about evolution in your high school biology course(s)?
    a) Yes, and creation was taught along with it
    b) Yes, and creation was NOT taught along with it
    c) No

29. Do you think the modern theory of evolution has a valid scientific foundation? (select one)
    a) Yes, because it is possible to test many hypotheses of evolutionary theory
    b) Yes, even though we can never test hypotheses about events in the past
    c) No, because we can never be sure about events in the past
    d) No, because evolutionary theory is based mainly on speculation, not hard scientific facts
    e) No, because it goes against my convictions

30. Which of the following best agrees with your conception of the modern theory of evolution? (select one)
    a) Man evolved from an apelike ancestor in Africa
    b) Evolution occurred because some varieties of organisms left more offspring than others
    c) Evolution involved a <u>purposeful</u>, steady progress by organisms from lower to higher forms
    d) Evolution occurred because the strong species eventually eliminated the weak species

**For Numbers 31–44,
please tell us how reliable a source
of information you consider each of the
following to be, using the following scale (select one):**

**a = Very Reliable b = Somewhat reliable
c = Unreliable d = I don't use this source**

31.	TV news programs	a	b	c	d
32.	Popular science magazines, like Discover or Science 94	a	b	c	d
33.	Daily newspapers	a	b	c	d
34.	National Enquirer, Star, or similar papers	a	b	c	d
35.	"In Search Of," "That's Incredible," or similar TV shows	a	b	c	d
36.	Books reporting on UFOs, the Bermuda Triangle, etc.	a	b	c	d
37.	Science programs on PBS like "Nova" or "Cosmos"	a	b	c	d
38.	Science fiction or fantasy books or stories	a	b	c	d
39.	Books/articles by science writers like Carl Sagan	a	b	c	d
40.	Professional scientific journals like Science or Nature	a	b	c	d
41.	Books advocating creationism by Duane Gish & others	a	b	c	d

42.	News magazines such as Time	a	b	c	d
43.	Your college instructors	a	b	c	d
44.	Your parents	a	b	c	d

PART II

Select the phrase after each statement that most clearly describes your belief about the statement. Possible choices are(select one):

a) Strongly agree

b) Agree

c) Undecided

d) Disagree

e) Strongly disagree

f) Never heard of it/don't know enough to have an opinion

45.	The earth is between 4 and 5 billion years old.	a	b	c	d	e	f
46.	Human beings came to be through evolution, which was controlled by God.	a	b	c	d	e	f
47.	There is intelligent life somewhere out there in the universe.	a	b	c	d	e	f
48.	Adam and Eve, the first human beings, were created by God.	a	b	c	d	e	f

49.	There is a good deal of scientific evidence against evolution and in favor of the Bible's account of Creation.	a	b	c	d	e	f
50.	UFO's are actual spacecraft from other planets.	a	b	c	d	e	f
51.	Science has done far more good than bad for the world.	a	b	c	d	e	f
52.	One can believe in the Bible and Creation, <u>or</u> in atheistic evolution; there is really no middle ground.	a	b	c	d	e	f
53.	Reincarnation really happens.	a	b	c	d	e	f
54.	Human beings, biologically identical to us today, have been around for at least 40,000 years.	a	b	c	d	e	f
55.	It is impossible to communicate with the dead.	a	b	c	d	e	f
56.	The theory of evolution correctly explains the development of life on earth.	a	b	c	d	e	f
57.	Everything written in the Bible is literally true.	a	b	c	d	e	f
58.	People <u>cannot</u> predict future events by using psychic power (ESP).	a	b	c	d	e	f
59.	God created humanity pretty much in its present form within the last 10,000 years or so.	a	b	c	d	e	f

		a	b	c	d	e	f
60.	The Biblical story of the Great Flood and Noah's Ark is symbolic rather than an actual event.	a	b	c	d	e	f
61.	Most scientists are atheists.	a	b	c	d	e	f
62.	Human beings came to be through evolution, which occurred without the help of God.	a	b	c	d	e	f
63.	Claims that there is some mysterious force operating in the Bermuda Triangle are untrue.	a	b	c	d	e	f
64.	Most scientists today believe that the modern theory of evolution is a valid scientific theory.	a	b	c	d	e	f
65.	The theory of evolution should be taught in public schools as an explanation of human origins.	a	b	c	d	e	f
66.	I have a clear understanding of the meaning of the scientific method of study.	a	b	c	d	e	f
67.	Science and religion often contradict each other.	a	b	c	d	e	f
68.	There is scientific evidence for the Biblical Great Flood.	a	b	c	d	e	f
69.	Humans evolved from other life forms solely through the laws of nature.	a	b	c	d	e	f

70.	Dinosaurs and humans lived at the same time, as shown by finds of their footprints together.	a	b	c	d	e	f
71.	The Bible's account of creation should be taught in public schools as an explanation of origins.	a	b	c	d	e	f
72.	Life began as a result of natural chemical reactions, and not through the power of a supernatural being or Divine intelligence.	a	b	c	d	e	f
73.	Life came to earth from other worlds in the universe.	a	b	c	d	e	f
74.	Psychics routinely predict future events and and help police solve crimes.	a	b	c	d	e	f
75.	God created "life" in its earliest form, then His natural laws took over to eventually produce all species, including human beings.	a	b	c	d	e	f
76.	Astrology is an accurate predictor of the future.	a	b	c	d	e	f
77.	Science makes our way of life change too fast.	a	b	c	d	e	f
78.	The first of every species (including humans) was created directly by God.	a	b	c	d	e	f

79. Angels really do exist, and can intervene in the lives of human beings. a b c d e f

80. One can believe in scientific evolution and the existence of God or a Supreme Being—one belief does not exclude the other. a b c d e f

81. It is not possible to read other people's thoughts using psychic powers (mental telepathy). a b c d e f

82. Science can explain human origins, but not the origin of the human soul or spirit a b c d e f

Index